Fermentation Kinetics and Modelling

C. G. Sinclair and B. Kristiansen

Edited by

J. D. BU' LOCK

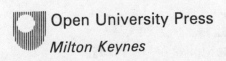

Open University Press
Milton Keynes

Taylor & Francis

New York • Philadelphia

Co-published by:
Open University Press
Open University Educational Enterprises Limited
12 Cofferidge Close
Stony Stratford
Milton Keynes MK11 1BY, England

Taylor and Francis
Publishing office   Sales office
3 East 44th Street   242 Cherry Street
New York NY 10017   Philadelphia, PA 19106
USA   USA

First Published 1987

British Library Cataloguing in Publication Data

Sinclair, C.G.
   Fermentation kinetics and modelling. —
   (Biotechnology series/Open University Press)
   1. Fermentation — mathematical models
   I. Title II. Kristiansen, Bjorn
   III. Bu'Lock, J.D. IV. Series
   660.2'8449'0724   TP156.F4

ISBN 0–335–15156–6
ISBN 0–335–15155–8 Pbk

Library of Congress Cataloguing in Publication Data

Sinclair, C, G. (Charles Gordon), 1929–
   Fermentation · kinetics and modelling.
   Bibliography: p.
   Includes index.
   1. Fermentation. I. Kristiansen, B. (Bjorn)
II. Bu'Lock, J. D. (John Desmond) III. Title.
TP156.F4S56 1987   660.2'8449'0724   87–1947

ISBN 0–8448–1509–8
ISBN 0–8448–1510–1 (pbk.)

Printed in Great Britain

Contents

FOREWORD : WHY MODEL ?

John D. Bu'Lock
University of Manchester

This book has been devised to deal with the construction and manipulation of mathematical models of fermentation processes, in the belief that making such techniques more accessible will be of general service to biotechnologists, whatever their technical background. My own introductory section is intended mainly for readers who are new to the subject; those who are already convinced about the value of kinetic models, and who in taking up this book simply want to be better equipped to use them in a microbiological context, can perhaps save time by proceeding directly to Chapter One.

First, what are the "fermentation processes", to which the modelling techniques are to be applied ? The key-word fermentation is not being used here in any narrow sense, but as applying quite generally to processes carried out by micro-organisms. It might also be applied equally well to processes carried out by plant and animal cells, whenever these are being handled - as is increasingly the case in modern biotechnology - by essentially the same techniques as are used for micro-organisms. Thus in general parlance the "fermentation industries" include not only brewing and other essentially anaerobic processes, but the whole repertoire of aerobic processes including the production of antibiotics, organic acids, vitamins, etc. as well as microbial biomass itself.

It would be quite inappropriate to attempt any account of all these different processes here; not only is the task too large, but in the present context we are seeking to develop a treatment that is sufficiently general not to require such details until it is applied to some specific situation. What all these processes have in common, however, is, first, the production, by cell multiplication in a suitable installation supplied with appropriate raw materials, of the necessary amounts of the desired micro-organisms, and, second, the provision - either during growth or as some separate operation - of the necessary raw materials and conditions that allow the micro-organisms to produce the desired products. Practical installations will also include arrangements for preparing the raw materials and for recovering the products, but these are outside our scope. What is important here is that if the process is to be an

economic one (and if it's not economic, it's not biotechnology) it has to be optimised in all these aspects, and it
is a characteristic of the way in which biological systems
respond to their environment that the seemingly separate
aspects will interact in quite complex ways to determine the
overall quantities of materials involved and the rates at
which they are formed or consumed. We shall need to develop
accounts of all this in describing our process.

Second, why will it be useful to be able to apply modelling
techniques to these processes ? To the laboratory microbiologist, some "fermentations" may appear very crude, as
for example the aerobic treatment of sewage with an
essentially adventitious microbial population. Equally
some, like the production of insulin by a genetically-
transformed bacterium, would appear highly sophisticated.
However, for their practical operation, both the "crude" and
the "sophisticated" processes are equally susceptible of
kinetic analysis. To describe them to someone, in anything
other than very specific terms, we shall want to say not
only what happened, but also how quickly it happened, and
how the results were affected by what we did, and how they
are likely to be affected by things we haven't got round to
doing yet. In short, we need a model; moreover any success
in modelling will be an essential step in developing new or
improved ways of carrying the process out in practice.

Even in the microbiology laboratory, we are aware that all
our experimental observations are necessarily carried out
under circumstances more limited than those to which we
would like our conclusions to be applicable, and we need
some principles we can use in order to generalise from what
we have observed. This is true even when our studies are
purely "observational"; the problem is more apparent with
biochemical or physiological investigations and it becomes
particularly obvious when we are hoping to develop an
industrial process. For industrial development, we will
necessarily carry out our investigations in set-ups which
will differ considerably from those we hope will eventually
be used.

*Whether we seek to interpolate "between" measured data, or
to extrapolate "beyond" them, or to apply them in a
generalised way and after due modification to radically
different set-ups, we will be using a model.*

Sometimes the model is so simple that we hardly notice it is
there. When we measure the optical density of a bacterial

culture at six-hourly intervals, and then "join up the dots" with a smooth curve, we are using a model - here a very simple one, which simply states that bacterial populations increase steadily and not in steps. We "know" that the model will be adequate if we are dealing with a sufficiently large number of bacteria under non-synchronized conditions, but also that it will break down over short intervals, when applied to a very few bacteria or to a synchronously-dividing population. So even this "implicit" model has to be used with care.

Though this book deals with more sophisticated models, the same cautions still apply, and are spelt out more carefully. The need to have such a model becomes more pressing as the system under study becomes more complex; at the same time, unless the model is really well-understood, it is always liable to be used wrongly.

Over the last fifty years, microbiologists have slowly become accustomed to using mathematical expressions to describe what they observe, even though there are some who still hanker after times when a microscope and a sketch-pad were all that was needed. Today terms like "log phase" and "specific rate" are common currency, but because of a lack of mathematical outlook they are often used loosely and, sometimes, quite wrongly. For instance, a substantial proportion of the papers in which the term "log phase" appears apply it wrongly, simply because the authors have never taken the trouble to plot their observations on log paper so see what is really happening. If you have built up the mathematical description yourself, you will be a lot more careful how you use it - and your eventual readers will have less excuse for misunderstanding what you say !

Indeed it can be argued that a model of a fermentation process that has been developed by a microbiologist, who has learnt the simple but necessary mathematical skills, is likely to be "better" than one developed from the same data by a mathematician, because the microbiologist has also, implicitly or explicitly, put into the model concepts and understanding drawn from his wider knowledge of how organisms behave. As a specific example, Sikyta* has argued that a model in which the growth kinetics of a microbial

---

*   B. Sikyta, M. Novak and P. Dobersky, Modelling of
    Microbiological Processes, in "Biotechnology and Fungal
    Differentiation", edited J. Meyrath and J.D.Bu'Lock,
    Academic Press (London) 1977, pp 119-136.

population are modelled by exponential functions is in this sense "better" than one using logistic functions or polynomial series, even though either approach can be used to "fit" the experimental data and will give expressions that can be further manipulated with comparable ease. This is simply because the mathematical terms in the exponential functions also correspond to a biological concept, the proliferation of daughter cells by the division of mother cells, while those in the polynomials do not.

In Chapter 1 of the present book, our authors emphasize that the development of a model is done for a purpose, and indeed that the nature of the aim will determine the character of a model. In the paper already cited, Sikyta has summarized the main uses of models as:
  (a) in experiment planning - the determination of limiting experimental conditions;
  (b) in interpolation and extrapolation of experimental results
  (c) for elucidating the nature of the process, its biological, chemical, or physical basis.
Each of these uses will impose specific requirements, depending on the nature of the experiments or their results, and the kind of understanding that the experimenter wishes or is equipped to reach. As is pointed out in Chapter 1, the individual experimenter may be interested in the process economics, or its power requirements, or in designing a new process that will operate at a thousand times the scale on which the experiments have been carried out. In each case the kind of experimental data needed or used will be quite different, and so will the nature of the model - it will be one that uses expressions that are meaningful to the economist or the engineer or the process microbiologist, as the case requires.

One important function of the model, however, should be to bring to attention factors which otherwise might be over-looked. This is the stage at which the engineer's model needs to be examined by the economist; the microbiologist's model requires scrutiny by the engineer. The engineer will soon learn that from the economist's point of view he cannot equate all forms of energy supply on a simple thermodynamic basis; the microbiologist will learn that phenomena like mixing times and temperature gradients, which he can ignore in his litre vessel, will have to be taken into account on the cubic metre scale. It is just this kind of dialogue that in the first place requires a model, to provide a basis for the discussion, and that in the second place demands that the mathematical terms in the model must be explained verbally. They must be "understood" - and it is precisely in this respect that some superficially very elegant models turn out to be relatively useless. For this reason I would urge any reader of this book *"keep your wits about you; think about every statement and try to illustrate it with things you already know"*.

For the reader trained exclusively in either of the necessary disciplines - basic microbiology or basic chemical engineering - there will be some problems of "jargon", *i.e.* of technical language. The authors have usually given explanations of special terms when they are first introduced, but to supplement this there is a glossary.

Apart from this, the only special difficulty a chemical engineer will find is the importance which is attached to *growth processes*. In most bio-processes, the catalyst on which the whole process depends is a population of living cells, whose number and whose specific activity will both vary with time and with the prevailing conditions. This means that kinetics which in inorganic systems are simply determined by the amount of catalyst put into the system, and the conditions arranged for its operation, are made more complex in the biological systems by the kinetics relating to the production (growth) and properties (physiology) of the catalyst itself. Moreover, these features are intrinsic to the whole process - they are not simply complicating arguments introduced by those fellows in white coats !

The classically-trained microbiologist will encounter rather different problems, needing to brush up elementary algebra and simple calculus, especially of rate processes, and - so far as possible - to try thinking (for just part of the time) in a more abstract way. This assignment may seem more difficult; in fact, working through this book should provide just what is needed. However, always beware of becoming too abstract ! In particular, beware of the spurious accuracy that comes from too much number-crunching; an equation is just as good as the experimental error in the data-points. Many arguments over which equation is "best" to describe this or that phenomenon would be avoided if their proponents only realised this; sometimes the only real conclusion is "it makes no difference" - or alternatively, "the experiment needs to be done better".

Happy modelling !

Manchester, March 1986

1.1 Introduction

A fermentation model is an abstracted and generalised description of relevant aspects of a fermentation process in which we are interested. Usually, but not inevitably, it is a mathematical model; often, though not always, it will be particularly concerned with the kinetic aspects of the fermentation. It is with the development, uses, and limitations of mathematical models of fermentation kinetics that this book is concerned.

The biotechnologist finds these models useful, or indeed quite essential, because he or she cannot investigate empirically the behaviour of any fermentation under all possible conditions and at all possible scales, but must be able to predict what the likely effects of changes of scale or conditions will be. Such predictions are quite essential if an operating process is being designed on the basis of laboratory investigations, but they are also useful in handling laboratory data, and in elaborating hypotheses to explain observations. In addition, the actual construction and checking of a kinetic model of a process is often the first step, and a very powerful aid, in designing further experiments, in exposing areas of unsuspected ignorance, and in furthering general understanding of a problem.

Students without a mathematical background may be discouraged from considering or using mathematical models by the apparent complexity of the mathematical equations which are their formal expression, and by the facility - often more apparent than real - with which many authors and lecturers appear to manipulate them; much of the published literature is confusing (to say the least) and by no means all of it is correct. In fact the work required to grasp the essential techniques of modelling is not excessive and the main steps the student should follow in writing models are all set out in section 1.2

Once the three groups of equations described there - balance equations, rate equations and thermodynamic equations - can be identified and understood, half the problem is overcome. The remainder of the problem is then sorted out by following through examples, by filling in the logical steps (which authors tend to omit) and by tackling some problem of interest to the student with the assistance of a skilled practitioner.

The other essential requirements for modelling are more practical than mathematical - clear pictorial diagrams of the microbial process and an unambiguous and easily-understood nomenclature; it is often helpful to substitute a familiar nomenclature for any unusual symbols used by an author. It is by no means essential for the concepts used in constructing a model to have any biological reality or meaning, but in practice models whose concepts do correspond to biological theory as generally understood are more readily used by workers with a biological background - and to that extent they are in fact more useful. A final requirement is that the dimensions of every term in an equation should be checked for consistency (this often exposes errors in formulation of the equations) and that the physical meaning of every term can be understood.

1.2 Formulating the Model

A model is a __set of relationships__ between the __variables of interest__ in the __system being studied__.

A __set of relationships__ may be in the form of equations, graphs, tables, or even, as for many experienced plant operators, an unexpressed set of cause/effect relationships which are a picture of the process and which determine his or her actions in controlling it. All such representations constitute the basic structure of every model.

The __variables of interest__ depend upon the use to which the model is to be put. For example, when confronted with a fermenter, a biotechnologist will be concerned with the feed rate, rate and mode of agitation, temperature, viability of the microorganism, etc.; an electrical engineer would be interested in the motor currents, voltages etc.; a mechanical engineer with the stresses and strains in the structure; an acountant with the costs of the daily inputs and outputs of material and energy. Each specialist will see the same equipment but will look at it in quite a different manner. The abstracted physical model selects from the real physical object those physical and geometrical properties which are of importance for the particular way in which the modeller wishes to treat the system.

The __system being studied__ must be defined in some detail; in biotechnology this is usually a reactor containing microorganisms, or the processes downstream of such a reactor where the product is separated into its constituents.

For our purposes, an abstracted physical model is defined as a region in space throughout which all the variables of interest (e.g. temperature, concentration, pH, dissolved oxygen) are uniform. This is called the 'control region' or 'control volume'.

The control region may be of constant volume, as is usually assumed for a chemostat or simple batch fermenter, or it may vary in size as for example in a fed batch fermenter. It may be finite as in the normal representation of a well mixed fermenter, or infinitesimal as in a tower fermenter (in which the concentrations of substrate and product vary continuously throughout the liquid space, so that only over an infinitesimal thin slice can they be considered uniform; such a reactor can most usefully be modelled as a series of infinitesimal control regions).

The boundaries of the control region may be:
a) phase boundaries across which no exchange takes place e.g. the walls of a containing vessel.
b) phase boundaries across which an exchange of mass or energy takes place, e.g. the bubble-liquid interface,
c) geometrically defined boundaries within one phase, across which exchanges take place either by bulk flow or by molecular diffusion, e.g. nutrient inlet and outlet pipes.

An example of an abstracted physical model for a fermenter, and how it relates to an actual fermenter, are shown together in Figure 1.1. Note how the abstracted model omits features which are not relevant to the matters of interest, and represents others in a purely formal way.

To construct a conventional mathematical model, we write a set of equations for each control region. This set consists of:-

A. **Balance equations** for each extensive property of the system i.e. mass, energy or individual elements or species. Where the extensive property is also a conserved property which can neither be created nor destroyed (such as mass, energy or chemical elements), then the balance equations are called conservation equations. ('Extensive property' is defined in section 2.1).

B. **Rate equations**; these are of two types:
  B.1   Rates of transfer of mass, energy, individual components or species <u>across</u> the boundaries of the region.
  B.2   Rates of generation or consumption of individual species <u>within</u> the control region

C. **Thermodynamic** equations which relate thermodynamic properties (pressure, temperature, density, concentration) either within the control region (e.g. gas laws) or on either side of a phase boundary (e.g. Henry's law).

Models are constructed to be used. The simpler they are, the easier they are to use, and so the golden rule of the model maker must be:-

<u>Always use the simplest adequate model firmly rooted in known fundamental physical, chemical and biochemical ideas</u>

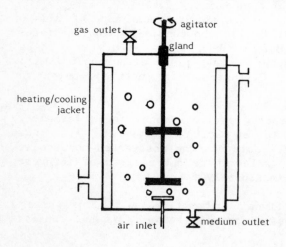

Simplified drawing of a fermenter

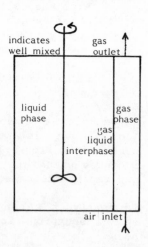

Abstracted physical model of a fermenter with the liquid phase as control region

FIGURE 1.1.

2.1 Balance Equations

Balance equations will need to be written for every extensive property of interest in each control region. The only
point to watch is that each balance equation is linearly
independent of the others.

Extensive properties are those which are additive over the
whole of a system, such that the amount of the property in
the whole system is the sum of the amounts in the separate
parts of the system. Thus mass or energy are extensive
properties, but temperature or concentration are not.

Linearly independent means that no balance equation can be
formed by adding together any combination of the others.

The balance equation is written:

rate of accumulation in control region	=	rate of input to control region	−	rate of output from control region

In writing down the balance equation it is often helpful if
the input and output terms are further subdivided, into:

Input terms: (i) bulk flow across geometrical boundaries

(ii) diffusion across geometrical boundaries
(only important for infinitesimal regions)

(iii) transfer across phase boundaries

(iv) generation within the control region

Output terms: (i) bulk flow across geometrical boundaries

(ii) diffusion across geometrical boundaries

(iii) transfer across phase boundaries

(iv) consumption within the control region

If generation is included in the input term and consumption
in the output, their signs are automatically correct.

2.2 Rate of Accumulation

Writing this term seems to give beginners more trouble than any other term in the balance equation. The first thing to remember is that it is not a kinetic, mass transfer or heat transfer rate term; it is simply the rate at which the amount of the extensive property within the control region changes with respect to time.

For example if x is the concentration of cells in a well mixed fermenter, in kg m^{-3}, and the volume of the control region is V m^3, then the amount of the extensive property 'mass of cells' is Vx (kg) and the rate of accumulation of cells in the control region is d(Vx)/dt (kg h^{-1}).

This formulation makes no assumptions about the constancy or otherwise of the control region volume. To express the rates as amount per unit volume then we must divide the term by V giving:

$$\text{Rate of accumulation} = \frac{1}{V}\frac{d(Vx)}{dt} \quad (\text{kg m}^{-3}\text{ h}^{-1})$$

2.3 Input and Output Terms

2.3.1. Bulk flow

This is an expression for material or energy carried by the bulk flow of fluid into or out of the control region and is thus equal to the flow rate (in m h^{-1}) multiplied by the concentration of the extensive property (in kg m^{-3} or J m^{-3}). If all the terms are to be based on unit volume then this quantity must be divided by the control region volume as before. Thus for example, taking the cell mass as the extensive property, if F is the medium input flow rate (in m^3 h^{-1}) then the bulk flow input term is Fx (in kg h^{-1}) based on total volume or Fx/V (in kg m^{-3} h^{-1}) based on unit volume of the control region.

2.3.2. Transfer across phase boundaries.

This is usually expressed on a unit volume basis and is then given by the product of three terms, a transfer coefficient, a phase boundary area term and a driving force term. The driving force term is the potential which causes the transferring species to move across the phase boundary; strictly this should be chemical potential for chemical species. In practice it is rather simpler if a concentration or partial pressure driving force term is used.

The transfer equation is therefore of the form:

| rate of transfer | = | transfer coefficient | x | area per unit volume | x driving force |

the units being, respectively:

$(kgm^{-3}h^{-1})$ $(kg \ m^{-2} \ h^{-1} \ DFunit^{-1})$ $(m^2 m^{-3})$ (DF unit)

The most common example of such an equation is that for the rate of mass transfer of oxygen in a fermenter, which is usually written:

$$N = k_L a \ (C_g^* - C_o) \tag{2.1}$$

On a total control volume basis both sides of the equation must be multiplied by the control region volume V, so that NV then has the units $(kg \ h^{-1})$; 2.1 is the more usual form.

2.3.3. Generation or consumption terms

These are always written as r with an appropriate subscript; for example r_x is the rate of growth of cells, r_p is the rate of production of some product, r_s is the rate of consumption of substrate, and so on. Usually, such rates are based on unit volume, and in each case the units will then be $kg \ m^{-3} \ h^{-1}$. To express the rates on a total control region volume basis each term must be multiplied by the volume.

The actual form of these expresssions are dealt with very briefly in the next section, and in rather more detail in chapter 3.

2.4 General Model for a Single Vessel

Consider an idealised fermentation process in which growing cells are consuming substrate, and producing more cells according to the following scheme.

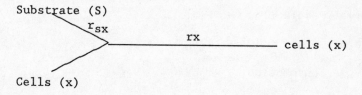

where:

r_{sx} =   rate of substrate consumption   kg m^{-3} h^{-1}

r_x =   rate of cell growth   kg m^{-3} h^{-1}

S   =   substrate concentration   kg m^{-3}

x   =   cell concentration   kg m^{-3}

To write a material balance for the extensive properties of interest in this example, i.e. the total amount of cells Vx and the total amount of substrate VS, where V is the volume of the control region, we will need the help of a pictorial diagram, as shown in Figure 2.1. Using the liquid in the fermenter as the control volume, the material balances can then be written as follows:

Rate of accumulation   =   rate of input   -   rate of output

<u>Cells:</u>

Accumulation: $= \dfrac{d(Vx)}{dt}$

Input terms: bulk flow $= F_i x_i$

                generation $= r_x V$

Output terms: bulk flow $= F_o x_o$

giving: $\dfrac{d(Vx)}{dt}$ $= F_i x_i + r_x V - F_o x_o$ (2.2)

<u>Substrate:</u>

Accumulation: $= \dfrac{d(VS)}{dt}$

Input terms: bulk flow $= F_i S_i$

Output terms: bulk flow $= F_o S_o$

                consumption $= r_{sx} V$

giving: $\dfrac{d(VS)}{dt}$ $= F_i S_i - F_o S_o - r_{sx} V$ (2.3)

In addition the change in reactor (liquid) volume can be accounted for by:

$$\frac{dV}{dt} = F_i - F_o \qquad (2.4)$$

Recall that the definition of the control volume states that the conditions are uniform throughout, i.e. the reactor is well mixed, and therefore in this case we can write:

$$x_o = x \qquad (2.5)$$

$$S_o = S \qquad (2.6)$$

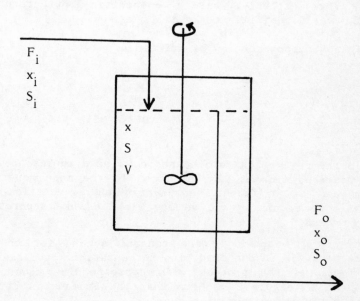

F   = flow rate $m^3\ h^{-1}$
x   = biomass concentration $kg\ m^{-3}\ h^{-1}$
S   = substrate concentration $kg\ m^{-3}\ h^{-1}$
i,o = inlet and outlet conditions respectively

FIGURE 2.1. Diagram of a simple fermentation process.

Considering also the simple case in which the liquid volume is constant ($F_o = F_i = F$), and supposing that there are no cells in the incoming medium (e.g. with a sterile feed), our original equations 2.2 and 2.3 now reduce to :

$$\frac{dx}{dt} = r_x - Dx \qquad (2.7)$$

and

$$\frac{dS}{dt} = -r_{sx} + D(S_i - S) \qquad (2.8)$$

where

$$D = \text{dilution rate} = F/V \quad (h^{-1})$$

Equations 2.7 and 2.8 should be familiar to the reader, as they are used extensively in the literature.

In their most common form, r_x and r_{sx} are defined as μx and $\mu x/Y$ respectively, where μ = specific growth rate (h^{-1}) and Y = yield coefficient (kg cells/kg substrate).

The equations can now be written as:

$$\frac{dx}{dt} = \mu x - Dx \qquad (2.7a)$$

and

$$\frac{dS}{dt} = -\frac{\mu x}{Y} + D(S_i - S) \qquad (2.8a)$$

In the derivations of the above we have represented a highly idealised fermentation process. In the more general case a number of different consumption and generation reactions will be occurring and we may wish to take separate account of these.

Figure 2.2 shows a more general scheme for the reaction mechanisms encountered in fermentation processes, and a sketch of the process with corresponding nomenclature is shown in Figure 2.3. Note that the biomass is here divided into viable and non-viable cells.

Material balances for all the extensive properties of interest (excluding lysed cell concentration) that is, for x_v, x_d, S and P can be written as described in section 2..1.

By assuming also that the maintenance energy term is provided by consumption of the limiting substrate, we get for viable cells:

Rate of accumulation $= \dfrac{d(Vx_v)}{dt}$

Inputs:

bulk flow $= F_i x_{vi}$

rate of cell growth $= Vr_x$

Outputs:

bulk flow $= F_o x_{vo}$

rate of formation of

non-viable cells $= -Vr_d$

Hence

$$\frac{d(Vx_v)}{dt} = F_i x_{vi} + Vr_x - F_o x_{vo} - Vr_d \qquad (2.9)$$

which is normally written as:

$$\frac{d(Vx_v)}{dt} = Vr_x - Vr_d + F_i x_{vi} - F_o x_{vo} \qquad (2.9a)$$

Similarly for the other variables we get:

non-viable cells:

$$\frac{d(Vx_d)}{dt} = Vr_d + F_i x_{di} - F_o x_{do} \qquad (2.10)$$

substrate:

$$\frac{d(VS)}{dt} = -V(r_{sx} + r_{sm} + r_{sp}) + F_i S_i - F_o S_o \qquad (2.11)$$

product:

$$\frac{d(VP)}{dt} = Vr_p + F_i P_i - F_o P_o \qquad (2.12)$$

total volume:

$$\frac{dV}{dt} = F_i - F_o \qquad (2.13)$$

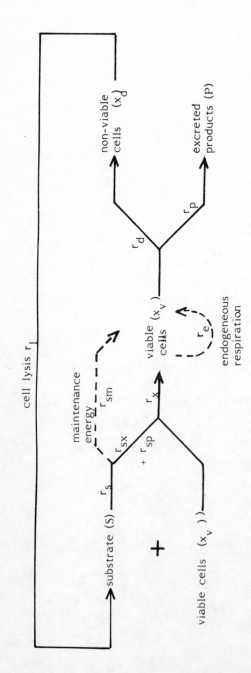

r_s = overall rate of substrate consumption (kg m^{-3} h^{-1}) ; r_{sx} = rate of substrate consumption for biomass production (kg m^{-3} h^{-1}) ; r_{sp} = rate of substrate consumption for product formation (kg m^{-3} h^{-1}) ; r_x = rate of cell growth (kg m^{-3} h^{-1}) ; r_{sm} = rate of substrate consumption for production og maintenance energy (kg m^{-3} h^{-1}) ; r_e = rate of consumption of organic cell matter in endegenous respiration (kg m^{-3} h^{-1}) ; r_p = rate of product formation (kg m^{-3} h^{-1}) ; r_d = rate of cell de-activation (cell death) (kg m^{-3} h^{-1}) ; r_l = rate of cell lysis (kg m^{-3} h^{-1}).

FIGURE 2.2. Basic biological reaction mechanism.

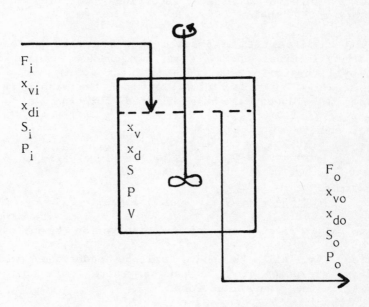

F = flow rate $\qquad$ m^3 h^{-1}

x_v = viable cell concentration $\qquad$ kg m^{-3}

x_d = non-viable cell concentration $\qquad$ kg m^{-3}

S = substrate concentration $\qquad$ kg m^{-3}

P = product concentration $\qquad$ kg m^{-3}

V = reactor (liquid) volume $\qquad$ m^3

i,o = inlet and outlet coditions respectively

FIGURE 2.3. General fermentation process for a single vessel.

The rate of consumption of substrate for the production of maintenance energy, r_{sm}, which is included in the material balance for the substrate, equation 2.11, is considered further in 3.3, below.

If alternatively it is assumed that the cell produces the required internal energy from endogeneous respiration and not by diversion of substrate (insofar as the two concepts are meaningfully distinguishable) then the viable cell and substrate balances must be modified as follows:

viable cells:
$$\frac{d(Vx_v)}{dt} = V(r_x - r_d - r_e) + F_i x_{vi} - F_o x v_{vo} \quad (2.14)$$

substrate:
$$\frac{d(VS)}{dt} = -V(r_{sx} + r_{sp}) + F_i S_i - F_o S_o \quad (2.15)$$

where r_e = rate of endogeneous respiration (kg kg^{-1} h^{-1}).

Notice that all the above balance equations, with the trivial exception of 2.13, can be written on a unit volume basis, in the generalised form:

$$\frac{1}{V}\frac{d(Vy)}{dt} = \sum r_{gen} - \sum r_{cons} + Dy_i - \gamma Dy_o \quad (2.16)$$

in which:

y stands for the general extensive property, in this case x_v, x_d, S or P

$\sum r_{gen}$ means the sum of all the rates of generation

$\sum r_{con}$ means the sum of all the consumption rates

(both in kg of y m^{-3} h^{-1})

$D = F_i/V$ and $\gamma = F_o/F_i$

In applying these forms, recall that the definition of the control volume states that the conditions within it are uniform throughout (i.e. the reactor is well mixed), thus:

$y = y_o$ throughout.   (2.17)

2.5 Structured and Unstructured Models

2.5.1 Unstructured mechanisms

Kinetic expressions have to be used or developed for the various steps in the overall biological process, and to do this it is necessary to agree on a basic biological mechanism . This may be at various levels of detail. For example it would in principle be possible to set out every one of the reaction steps that occur within the cell and then to write the overall set of equations to describe all of them. Quite apart from it being an impractical task, such a model would be far too detailed for technological purposes, though (of course in a simplified form) it might be of interest to biochemists. A less detailed mechanism might identify one or two key chemical species within the cell, perhaps those which were to be extracted on a commercial scale, and lump all other components of the cell under one or two general classes such as proteins, RNA, lipids etc. This could be done on the basis of what is known about their role and mode of formation.

Such mechanisms are the basis of **structured** models, i.e. those which take into account some basic aspects of cell structure, function and composition. However, what can be termed an **unstructured** mechanism of cell operation is sufficient for many technological purposes, and will be sufficiently simple to give rise to sets of equations which can be understood in a physical sense while still having sufficient biological significance to be generally helpful.

In an unstructured mechanism of cellular operation, the microorganism is regarded as a single reacting species, possibly with a fixed chemical composition, and the basic biological "reaction scheme" is the one we have already shown in Figure 2.2. Here we identified substrate, viable cells, non-viable cells and excreted products as the four major components. Substrate reacts with viable cells to form more viable cells, which may be harvested, may excrete products into the medium, or may be converted into non-viable forms of the cell. The non-viable cells may then form part of the harvested cell mass or may themselves lyse to release their constituents into the medium.

The energy necessary for growth and product formation is obtained either by breakdown of some of the substrate or by breakdown of cell material itself. In addition to the energy

required for synthetic reactions, energy is required to
maintain the cell in a viable state. As shown in Figure 2.2,
this requirement is termed maintenance energy if it is
assumed to come from the substrate, or endogeneous
respiration if we consider it to come from breakdown of cell
constituents (see also section 3.3).   The energy surplus
appears as evolved heat.

In Figure 2.2, each path is labelled with a subscripted
letter 'r', to designate the rates at which the labelled
reaction proceeds (in units of 'amount of reactant consumed
or product formed per hour per unit volume').   The subscript
can be mnemonically related to the particular reaction step.

Most of these rates involve sets of processes too complex to
handle numerically even if they were known in detail.   To be
able to handle these more easily, the idea of the limiting
substrate or substrates is invoked, (see section 3.2). Thus
out of all the substrate constituents necessary for growth,
it is assumed that all but one (or at most two) are in
excess and that it is the concentration of these one or two
'limiting' substrates which controls the overall rate of
reaction.   When all the substrates are in excess (as at the
beginning of a batch culture), it is assumed that the cells
grow at some maximum rate determined by their intrinsic
nature and by environmental conditions other than soluble
substrate concentrations.

2.5.2 Structured mechanisms
The unstructured type of model discussed in the previous
section has played the major part in the development of our
understanding of fermentation processes; it is still the
basis of the technologist's view of biological processes and
of much research. Its limitations are that it does not make
proper use of our considerable knowledge of the processes
which occur within the cell, or of our ability to analyse
cells for particular constituents. Consequently although
models based on an unstructured mechanism are adequate for
interpolation between experimental results on a particular
fermenter, they may not allow for extrapolation either to
larger scales of operation or to radically different
environmental conditions.

It is therefore necessary to look at the cellular processes
in some detail. A normal procedure is to consider some
individual constituents of the cell and to group others into
a small number of general classes. One of the earliest and
simplest of these mechanisms for cell growth is that due to

Williams.   Figure 2.4   shows the mechanism in diagrammatic form. In this the circled letter 'S' represents the limiting substrate, 'D' the synthetic apparatus of the cell (i.e. its enzymes) and 'G' the genetic part.   Thus the cell is here considered to consist of two constituents only, called D-mass and G-mass.   The solid lines in this figure represent the reaction steps, and the broken lines the control over the reaction rates exerted by the fractional concentrations of the D-mass and G-mass.   In the model which Williams derived using this structured mechanism, he assumed that the fraction of G mass could only lie between two limits:

- a lower limit below which the cell is dead,

- an upper limit at which it divides.

His model was found to be useful in predicting the lag phase in batch culture. By assigning part of the D-mass to product formation and part to cell growth, the mechanism can also be adapted to describe product formation.

Another kind of structured mechanism which is sometimes useful is one which considers the ATP in the cell as a separate component.   We can then write a conservation equation with the principal mechanisms for consumption and generation of ATP as the main terms. This approach organises a large amount of biochemical information which is available on energy turnover in the cell and has proved very useful in such diverse problems as modelling alcohol production, citric acid formation, polysaccharide formation etc. More information on structured models is available in the literature; this text is confined to unstructured models.

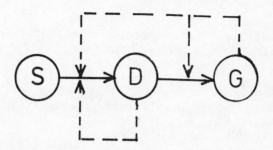

FIGURE 2.4. Williams' model for microbial growth in which the total biomass is divided into G-mass and D-mass. S is the substrate.

3 *Rate Equations*

3.1 General Principles

This section is confined to consideration of the kinetic expressions which relate the rates of generation or consumption of a chemical or biological species to the concentrations of nutrients etc. in the environment, and to the physical state of that environment. The basic form of the kinetic rate expressions, as used in the balance equation, will be either:

-   a specific rate on a cell mass basis, multiplied by the concentration of cells in the control region,

-   or a stoichiometric constant multiplied by a rate expression,

-   or a linear combination of these two.

An example of the first would be the well known expression for cell growth:

$$r_x = \mu x$$

where μ is the specific growth rate of cells, in kg cells per kg cells per hour (i.e., in h^{-1}) and x is the cell concentration in kg m^{-3}.

An example of the second is a common expression for growth-related product formation:

$$r_p = \alpha r_x$$

In this α is the stoichiometric coefficient, kg product per kg cells.

An example of the third (combined) type is the Luedeking and Piret expression for product formation:

$$r_p = \alpha r_x + \beta x$$

where β is the specific rate coefficient, kg product per kg cells per hour.

To derive relationships between substrate utilization, growth, and other cell functions, we need to consider their biological basis, at least in a generalised manner. This is an important aspect of model construction and one where mathematical concepts need to be understandable in

biological terms; equally it is one where the attempt to
derive mathematical statements can often clarify the
biological concepts.

Microorganisms require substrates for three main functions:

1.   to synthesise new cell material

2.   to synthesise extracellular products

3.   to provide the energy necessary
     a) to drive the synthetic reactions
     b) to maintain concentrations of materials
        within the cells which differ from those
        in the environment,
     c) to drive recycling (turnover) reactions
        within the cell

Thus growth, substrate utilisation, maintenance and product
formation are all intimately related, and as will be shown
later the various rate expressions are also mathematically
related.

The energy required to drive the cell processes is the
chemical energy of ATP or similar substances, which in most
fermentation processes is provided either aerobically, by
the oxidation of substrate by molecular oxygen to CO_2 and
water (oxidative phosphorylation) or anaerobically, by the
degradation of substrate to simpler products such as
ethanol, lactic acid, CO_2 and water etc., which are excreted
by the cell (substrate level phosphorylation). The products
of substrate level phosphorylation are called **type 1**
products.

Extracellular products will include such compounds as
   - exoenzymes (for breaking down substrates which cannot
     pass through the cell wall).
   - polysaccharides (for cell aggregation)
   - special metabolites (e.g. antibiotics; whose function
     may be to inhibit competing microorganisms but in the
     general case is unknown).
These products are called **type 2**.

It may also be useful to distinguish **type 3** products; these
are substances produced in situations where the carbon
substrate is in excess and other substrates such as nitrogen
or magnesium are limiting. They include possible 'energy
storage' compounds such as glycogen or fat etc. which are

stored within the cell, or similar polysaccharides etc. excreted by the cell. These type 3 products are considered by some to act as an 'energy sink'; excess ATP is produced so as to use the limiting substrate more efficiently, and the formation of type 3 products then dissipates the chemical energy of the excess.

The total energy required to maintain the concentration gradients which usually exist between the interior and the exterior of cells, and to drive turnover reactions in which labile cell components are continuously re-synthesized [i.e. for functions 3 (b) and 3 (c) above] is usually referred to as the maintenance energy requirement, being used only to maintain the cell in a viable state and not to produce cell material or products of types 2 or 3.

The general relationship between the flows of cells, substrates, energy and products is represented in Figure 3.1, showing the processes which occur in the cell diagrammatically. The solid lines represent material flows and the broken lines energy flows in the form of ATP. Carbon substrate (represented as C) forms endogeneous products type 3 and with nitrogen substrate (N), combines to form cell mass and type 2 products. All of these processes require energy. Carbon substrate is broken down, either anaerobically to type 1 products or aerobically to CO_2 and water, to provide energy for the previously mentioned syntheses and for maintenance. Each of these streams is labelled with its rate.

In general, a set of independent mathematical equations will have unique solutions if the number of unknown variables is equal to the number of equations that relate them (e.g. to solve for two unknowns, we need two independent equations relating them, and so on). Since in the present case, figure 3.1, we have specified three entering material streams (C, N, O), we can write three independent material balances, and we can also write the internal ATP balance; thus we will have four independent equations. Having specified eight material flows and one energy flow, that is nine rates in all, we can independently specify the kinetic expressions for any five of those nine rates. The other four kinetic expressions must then be related to these specified five.

It is normal to specify independent kinetic expressions for r_m, the maintenance energy requirement and for r_x, the growth rate. Other kinetic expressions which might be specified are those for the rate of oxygen uptake r_o, and for production of type 2 and type 3 products, r_{p2} and r_{p3}.

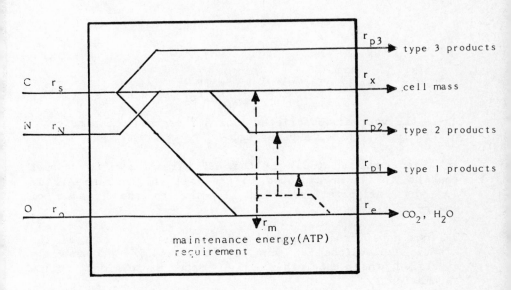

FIGURE 3.1. Cell processes.

r_N = rate of nitrogen consumption

r_s = rate of carbon substrate consumption

r_o = rate of oxygeb consumption

r_{p1} = rate of type 1 product formation

r_{p2} = rate of type 2 product formation

r_{p3} = rate of type 3 product formation

r_e = rate od endogeneous respiration

r_x = rate of cell growth

r_m = rate of maintenance energy production

In practice it is not possible to distinguish the CO_2 and water resulting from oxidative phosphorylation, r_c, from the CO_2 which is a type 1 product, and therefore both this flow and the corresponding oxygen balance are omitted from the analysis. The two material balances and the one ATP balance are therefore:

$$r_s = r_{sx} + r_{sp1} + r_{sp2} + r_{sp3} + r_{so} \qquad (3.1)$$

$$r_N = r_{Nx} + r_{Np2} \qquad (3.2)$$

$$a_x r_x + a_{p2} r_{p2} + a_{p3} r_{p3} + r_{mATP}$$
$$= a_{p1} r_{p1} + a_o r_o \qquad (3.3)$$

where the various subscripted a in 3.3 are stoichiometric coefficients, in mol ATP per kg.

The last equation (3.3) is the ATP balance and it assumes that no ATP is accumulated in the cell and that no ATP is lost by "slippage" reactions leading to the release of energy as heat.

We thus have three equations and seven rates so that if the kinetic expressions for four of the rates are independently specified then the other three rate expressions follow automatically.

The independently specified kinetic rate expression relates the rate of consumption of a substrate or the rate of formation of a product (including cell mass) to the concentrations of the various components in the environment of the cell and the other environmental conditions such as pH, temperature etc. Such an equation will have the general form:

$$r = f(x; S_1, S_2, S_3, \ldots; C_1, C_2, C_3, \ldots) \qquad (3.4)$$

where r is the rate (usually in amount per unit time per unit volume of medium)

    x is the concentration of microorganism in amount per unit volume

    S_1, S_2, etc are the concentrations of the various substrates in the medium in amount per unit volume

    C_1, C_2, etc are the other environmental variables.

The function f has to have some explicit form, and it is one of the objects of research to discover these forms for specific cases. Several possible expressions for the growth rate are given in the following section.

3.2 Cell Growth and Inhibition

3.2.1 Monod kinetics

Many of the kinetic models for cell growth and inhibition are based on those used in enzyme kinetics. The rationale behind this is the accepted picture of a cell as a miniature chemical reactor in which a complex network of enzyme catalysed reactions converts substrates into living cell matter and externally excreted biochemicals. Part of such a network is represented in Figure 3.2.

It is assumed that all but one of the substrates are available in excess of the cells capacity for absorption, whereas this one substrate is limiting. Again, of all the various routes by which this substrate is incorporated into the cell mass, it is assumed that one route is the slowest and thus will limit the overall reaction rate. Lastly within the limiting path, we presume that there is one reaction step which governs the overall rate, so that this is the limiting reaction step. Hence in this simplified model the overall rate of growth of the cell depends on this one enzymic reaction step and therefore on the effect of the concentration of the limiting substrate on the rate of that step.

The simplest of the expressions relating enzyme reaction rate to the limiting substrate concentration is the Michaelis-Menten expression,

$$v = \frac{k \, E \, S}{k_m + S} \qquad (3.5)$$

in which v is the reaction velocity

k is the rate constant

E is the total amount of enzyme

k_m is the Michaelis constant

and S is the substrate concentration

The product kE in (3.5) is the maximum rate at which the reaction can proceed (when $S \gg k_m$) and is often written as v_m.

If we identify an enzyme reaction of this type with the rate controlling step, and if we assume moreover that the concentration of this rate controlling enzyme is proportional to the cell concentration, while the concentration of the substrate for the rate controlling step is proportional to the limiting substrate concentration, then we can write an analogous expression, - the classical Monod equation for cell growth - as:

$$r_x = \frac{\mu_m\, S}{k_s + S}\, x_v \qquad (3.6)$$

where

r_x is the rate of cell growth, kg cells m^{-3} h^{-1}.

μ_m is the maximum specific growth rate, h^{-1}.

S is limiting substrate concentration, kg substrate m^{-3}.

k_s is the saturation constant, kg substrate m^{-3}.

x_v is the viable cell concentration, kg cells m^{-3}.

This expression is often written:

$$r_x = \mu(S).x \qquad (3.7)$$

where $\mu(S) = \mu_m S/(k_s + S)$ $\qquad (3.8)$

Normally the specific growth rate function $\mu(S)$ is simply abbreviated as μ, and it has the dimensions h^{-1}.

The various properties of this expression are illustrated in Figure 3.3. In particular, note (and confirm by examining the figure) that:

-    when the substrate concentration is not limiting, that is when $S \gg k_s$, the specific growth rate $\mu(S)$ approaches μ_m and the growth rate r_v becomes independent of S and simply proportional to the cell concentration x_v.

-    when the substrate concentration S is numerically

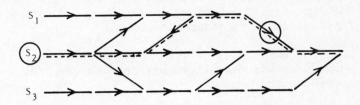

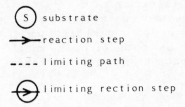

FIGURE 3.2. Schematic diagram of reaction network inside a cell.

(S) substrate

⟶ reaction step

---- limiting path

⊖→ limiting rection step

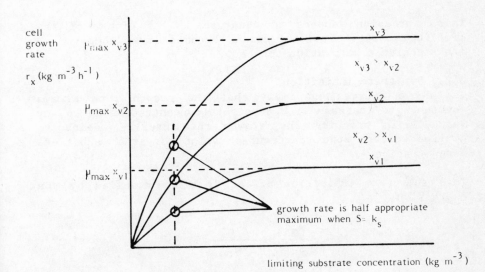

FIGURE 3.3. A plot of cell growth rate against limiting substrate concentration for Monod kinetics.

equal to the saturation constant, k_s , the specific growth rate $\mu = \mu_m/2$; for this reason k_s is often termed the critical substrate concentration.

when S is small compared with k_s, the specific growth rate is nearly proportional to the limiting substrate concentration.

Variants of equation 3.6 have been devised which are more often found useful in specific situations and some of these are listed below.

3.2.2 Double substrate limitation
For the case where there are two substrates S_1 and S_2 which are both present in such low concentrations that the cell growth rate is limited by both (i.e. a small increase in either substrate concentration will increase the growth rate), we can write:

$$\mu = \mu_m \left\{ \frac{S_1}{k_{s1} + S_1} \right\} \left\{ \frac{S_2}{k_{s2} + S_2} \right\} \qquad (3.9)$$

This expression reduces to equation 3.8 if either of the limiting substrate concentrations becomes large compared to its associated saturation constant.

3.2.3 Substrate inhibition
The above expressions imply that the growth rate always continues to increase with substrate concentration up to any value. In practice the growth rate usually begins to decline above some particular value of the substrate concentration.

This substrate inhibition effect is often modelled by the expression:

$$\mu = \frac{\mu_m S}{k_s + S + (S/k_i)^2} \qquad (3.10)$$

where k_i = inhibition coefficient.

Other expressions that can be used to describe kinetic situations including an inhibition term are described below.

3.2.4 Growth inhibition

There is a wide range of substances which inhibit growth, and the effects of increasing concentrations of these inhibitors mirror the effects of enzyme reaction inhibitors. Therefore all of the expressions that have been developed for enzyme inhibition can be applied to model cell growth inhibition. Amongst these are:

$$\mu = \frac{\mu_m \, S}{k_s + S} (1 - k_i I) \qquad (3.11)$$

$$\mu = \frac{\mu_m S}{k_s + S} \cdot \frac{k_i}{k_i + I} \qquad (3.12)$$

$$\mu = \frac{\mu_m S}{k_s + S} \exp(-k_i I) \qquad (3.13)$$

In these expressions:

$\qquad$ I $\quad=\quad$ inhibitor concentration

and k_i $\quad=\quad$ inhibition constant

Note that the coefficients k_i do not have the same meanings (or units) in the different expressions. The choice between them is partly a matter of convenience (one equation may be easier to manipulate than another) and partly requires reference to actual observations, made with sufficient accuracy over a sufficiently wide range of values of the inhibitor concentration. When the observations are not very accurate, or when only a limited range of data is available, there is little to choose between these (and other) expressions - as the reader may quickly verify by expanding 3.12 and 3.13 in series form.

There have been attempts to include these effects in the kinetic expression for cell growth rather than modelling them as separate phenomena, as for instance by Contois, who proposed

$$\mu = \mu_m S/(Bx + S)$$

where B is a coefficient, kg substrate per kg biomass.

In general it is important to remember basic modelling principles - use the expression which fits closest to the observed facts and is also intelligible in terms of physical and biological thinking.

3.3 Maintenance and Endogeneous Respiration

3.3.1 Maintenance energy is that part of the energy requirements of the cell that is used to maintain the cell in a viable state, for example for resynthesis of cell constituents which are continuously being degraded (turnover), and for maintaining concentration gradients between the interior and exterior of the cell (osmotic work). The rate of consumption of substrate to provide the energy for maintenance is written r_{sm} (kg substrate $m^{-3}h^{-1}$), and the only generally accepted kinetic expression for the maintenance energy requirement is:

$$r_{sm} = m_s x_v \qquad (3.14)$$

where m_s is the rate constant (kg subsrtate kg cells^{-1} h^{-1})

We can alternatively write the maintenance equation in terms of ATP as:

$$r_{mATP} = m_{ATP} x_v \qquad (3.15)$$

where r_{mATP} is the ATP requirement (mols ATP m^{-3} h^{-1})

and m_{ATP} is the rate constant (mols ATP kg cells^{-1} h^{-1})

The value of the rate constant m_s can range from as little as 0.02 to as high as 4 kg substrate kg cells^{-1} h^{-1} , and values of m_{ATP} range from 0.5 to over 200. Clearly maintenance energy can be a substantial fraction of the total energy consumption.

The value of the maintenance energy coefficient, m_s, will depend upon the environmental conditions surrounding the cell and on its rate of growth. A large part of the maintenance energy is required for osmotic work; thus increasing the external salt concentration increases m_s substantially, and pH also has a pronounced effect. The most rapid turnover of cell protein occurs when the cell is adapting its enzyme spectrum to changing environmental conditions. Under constant growth conditions the amount of maintenance energy for turnover is low but when growth stops

or when the cell is adapting to changes in substrate or to
cessation of growth then this becomes more significant. So
far there has been little quantification of these effects
and there are no kinetic expressions for m_s as a function of
medium composition, that would be comparable to the kinetic
expressions for cell growth.

3.3.2 Endogeneous respiration is an alternative way of
looking at the provision of energy for maintenance of cell
viability. It is assumed that the energy for maintenance is
provided by the oxidation or degradation of some of the cell
mass itself. This approach often creates conceptual
difficulties. It is clearly comprehensible when there is no
other source of energy, as when the external substrate
supply has run out, but less so when there is still an
excess of external energy-providing substrate. The rate
expression is written:

$$r_e = k_e x_v \qquad (3.16)$$

where

r_e is rate of endogeneous respiration (kg cell m^{-3} h^{-1})

and k_e is the rate constant (kg cell matter kg cells^{-1} h^{-1}).

It should be apparent that, according to whether we use the
maintenance energy concept or the endogeneous respiration
concept, the two consumption terms r_{sm} and r_e will appear in
different balance equations.

If we use the maintenance concept then r_{sm} appears in the
substrate balance, whereas if we use the endogeneous
respiration concept then r_e appears in the cell mass
balance.

However there is no practical difference between the
resulting sets of equations; we can convert from one to the
other by making the substitution $k_e = m_s Y_{x/s}$, where $Y_{x/s}$ is
the yield of biomass on substrate, except in the case where
the external substrate supply is less than the maintenance
requirement. In these circumstances the maintenance concept
breaks down, since it predicts a simple cessation of growth,
whereas the endogeneous concept predicts what is also
observed experimentally, namely a decline in cell mass with
time. More complex models involving a switch in the energy
supplying mechanisms according to external conditions would
eventually reconcile the two approaches.

3.4 Cell Death

There seems to be a natural rate at which cells become
non-viable i.e. incapable of growth and reproduction.
Sometimes the cells are truly dead and can only lyse, while
in other cases the cells enter a state of "suspended
animation" but do not revert to the viable state in times of
interest in commercial biotechnological processes. Thus the
word "viable" is being used here in a sense slightly
different from the way microbiologists use it. To handle
this in kinetic terms, the rate of conversion to the
non-viable form is assumed to be directly proportional to
the mass of viable cells and is written.

$$r_d = k_d x_v \qquad (3.17)$$

where

r is rate of conversion to non-viable form (kg cells $m^{-3}h$-1)

x is the concentration of viable cells (kg cells m^{-3})

k is the rate constant (kg cells kg cells^{-1} h^{-1})

Little is known of the influence of environmental conditions
on the rate constant k_d .

3.5 Product Formation

3.5.1 Classification
We have already partially classified products in section 3.1
on the basis of their relationship to the cell processes and
their appearance as flows of material out of the cell under
normal conditions. This approach to product classification
will now be extended.

Product classification

Type 1   Products of energy metabolism, i.e. by-products of
          the basic energy production processes in the cell.

Type 2   Extracellular products released by the cell

Type 3   Energy storage compounds

Type 4   Cell constituents. These are all the compounds
          present in the whole cells; they may conveniently

be divided into two sub classes,

Type 4a those which have to be recovered from the harvested cells by disruption of the cell.

Type 4b those which leak into the environment of the cell because of excessive build up within the cell or because of defects in the cell wall.

Many other classification schemes have been proposed in the literature. One of the earliest and more useful is due to Gaden, who related product formation to energy metabolism, distinguishing between three broad groups of products:

I   Products which are the result of primary energy metabolism, similar to type 1 above,

II   Products which arise indirectly from energy metabolism (such as intermediate metabolites),

III   Complex molecules not resulting directly from energy metabolism, similar to type 2 above.

This product classification has also led to a useful kinetic classification based on simple kinetic models which has been found to be of practical use. Each class is associated with a simple rate equation as follows:

I   Growth-associated $\quad r_p = \alpha r_x$ $\hspace{3cm}$ (3.18)

II Mixed kinetics $\quad\quad r_p = \alpha r_x + \beta x_v$ $\hspace{1.5cm}$ (3.19)

III Non-growth associated $\quad r_p = \beta x_v$ $\hspace{2cm}$ (3.20)

3.5.2 Kinetic models of product formation

A simple kinetic model for product formation is that which is the basis of the kinetic classification of products and was suggested by Luedeking and Piret for a lactic acid fermentation:

$$r_p = \alpha r_x + \beta x_v$$

which using equation 3.6 for cell growth expands to:

$$r = \alpha \left\{ \frac{\mu_m S}{k_s + S} + \beta \right\} x \hspace{1.5cm} (3.21)$$

This expression was originally suggested as an empirical model (i.e. one which fits the experimental data but is not based on any theoretical principles), but it can be derived from our basic picture of the cell process (section 3.1) as follows:

The ATP balance for a type 1 product such as lactic acid (for which $r_{p2} = r_{p3} = 0$) which is formed under anaerobic conditions ($r_o = 0$) is:

$$a_x r_x + r_{mATP} = a_{p1} r_{p1} \qquad (3.22)$$

but since $r_{mATP} = m_{ATP} x_v$ (from equation 3.15 above), then:

$$r_{p1} = \frac{a_x}{a_{p1}} r_x + \frac{m_{ATP}}{a_{p1}} x_v \qquad (3.23)$$

Thus the two terms of the Luedeking-Piret equation (3.21) correspond respectively to product formation linked to growth and to maintenance.

This theoretically based version of equation 3.19 suggests that for high rates of type 1 product formation under these conditions we need an organism with a high maintenance requirement (m_{ATP}) and a low production of ATP per unit of type 1 product produced (a_{p1}).

This derivation of the equation similarly explains why the bacterium *Zymomonas mobilis* is a more efficient producer of ethanol, another type 1 product, than is the yeast *Saccharomyces cerevisiae*. Firstly, the bacterium has a higher maintenance requirement; reported values of m_{ATP} for the bacterium are between 14 and 25 times the corresponding value for yeast. Secondly the bacterium uses a less "efficient" pathway of sugar breakdown to generate ATP; a_{p1} for the bacterium is 0.5 kmol ATP per kg ethanol produced, whereas for the yeast the value is 1.

The production of other products can be approached in a similar manner. For example for type 2 product formation we obtain from equation 3.3:

$$r_{p2} = \frac{a_o}{a_{p2}} r_o - \frac{m_{ATP}}{a_{p2}} x_v - \frac{a_x}{a_{p2}} r_x \qquad (3.24)$$

We can interpret this equation qualitatively as implying that for maximum product formation rate, cell growth should be suppressed (third term on right) and oxygen uptake rate maximised (first term on right). Substrate should be provided only for maintenance and for product formation according to the following equation:

$$r_s = \left\{ \frac{1}{Y_{p2/s}} + \frac{a_{p2}}{a_o} \frac{1}{Y_{o/s}} \right\} r_{p2} + \frac{m_{ATP}}{a_o} \frac{x_v}{Y_{o/s}} \qquad (3.25)$$

This expression can be derived by substituting for r_o from equation 3.3 into 3.1 using:

$$r_x = r_{p1} = r_{p3} = 0$$

$$r_{sp2} = \frac{r_{p2}}{Y_{p2/s}} \quad \text{and} \quad r_{so} = \frac{r_o}{Y_{o/s}}$$

The formulation of these stoichiometric relationships between rates is discussed further in section 3.6.

Kinetic models for product formation can also be adapted from the structured mechanisms for cell growth which were discussed in section 2.4. Little work has been done in this area, but it is one which is likely to become of increasing importance in the next few years. The general appoach is to take some measureable constituent of the cell, such as for example the RNA which has a product synthesis capability, and write rate equations describing the rate of production and degradation of this particular cell constituent.

The model might then be extended to cover (for example) enzyme production, by relating this to RNA concentration by means of a Monod type relationship (with RNA concentration replacing cell concentration and precursor concentration replacing the normal substrate). More complex models would be needed to relate the formation of the RNA to inducer compounds present in the medium. Thus although the individual rate equations would be simple, the overall model with all its ramifications can become quite complex.

3.5.3 Product inhibition and degradation
Products which reach a sufficiently high concentration in the medium may inhibit and eventually stop their own production. This phenomenon is well known in ethanol

production, where it limits the maximum strength of alcoholic beverages that can be produced by fermentation. The kinetic expressions for such inhibition will mirror the expressions for cell growth inhibition given in section 3.2. Indeed many products are also inhibitors of cell growth and it is only necessary to replace the inhibitor concentration I in these expressions by the product concentration P.

Thus a very commonly-used expression for modelling the inhibition of ethanol production is obtained from equation 3.19 by introducing a product inhibition term, giving

$$r_p = (1 - P/P_m).(\alpha r_x + \beta x_v) \tag{3.26}$$

where P_m is the maximum attainable ethanol concentration.

Products may be removed from the medium either by degradation, usually enzymatic, or by their utilization as substrate by the cells when their normal substrate becomes depleted. The consumption of a product as a substrate follows the normal kinetics for substrate uptake, while in the absence of more detailed knowledge, degradation of product is usually assumed to be a first order process.

3.6 Substrate Utilization

Substrate is used to form cell material and metabolic products as discussed in section 3.1, and the rate of substrate utilization is related stoichiometrically to the rates of formation of these materials. In some cases it is possible to write the chemical equations; for example the equation for ethanol production from glucose is:

$$C_6H_{12}O_6 = 2 C_2H_5OH + 2 CO_2$$

from which it is easy to calculate that 0.51 kg of ethanol will be formed from 1 kg of glucose. If for this case we write r_p as the rate of product (ethanol) formation and r_{sp} as the rate of substrate uptake for product formation then:

$$r_{sp} = r_p / 0.51 \tag{3.27}$$

More generally

$$r_s = \frac{r_p}{Y_{p/s}} \tag{3.28}$$

where $Y_{p/s}$ is the yield coefficient for product on substrate and has the units kg product formed per kg substrate converted to product.

From generalised observation, anaerobic cell growth might be described by:

$$CH_2O + 0.2\ NH_4^+ + e^- = CH_{1.8}O_{0.5}N_{0.2} + 0.5\ H_2O$$

where CH_2O represents the carbohydrate substrate, NH_4^+ is the nitrogen substrate, and $CH_{1.8}O_{0.5}N_{0.2}$ is an approximate formula for many cells (phosphorous and other elements which occur in much smaller proportions have been ignored).

From this that we can readily calculate that 0.82 kg of cells will be formed per kg of carbohydrate substrate utilised for cell formation, and 8.8 kg of cells per kg of nitrogen. Thus we could write as above:

$$r_{sx} = r_x / 0.82 \text{ or } = r_x / Y_{x/s} \qquad (3.29)$$

and $$r_{Nx} = r_x / 8.8 \text{ or } = r_x / Y_{x/N} \qquad (3.30)$$

where

r_{sx} is rate of carbohydrate utilization for cell growth

r_{Nx} is rate of nitrogen utilisation for cell growth

r_x is the rate of cell growth

$Y_{x/s}$ is the yield coefficient for cells on carbohydrate

$Y_{x/N}$ is the yield coefficient for cells on nitrogen.

It is important not to confuse these yield coefficients with "yields" as normally reported in the literature.

A yield coefficient is a stoichiometric constant which depends on the chemical equation relating the reactants and the products, whereas a yield is a ratio of one product to one reactant which may be entering into multiple reactions which form a variety of products. Thus in a normal fermentation process, with cell growth and product formation both consuming substrate, the "yield" will be calculated on the total amount of substrate consumed. We can illustrate this by a specific example.

In the anaerobic yeast fermentation, the energy required for synthesis is provided by the ethanol-forming reaction.

Using the nomenclature of section 3.1, we can calculate the cell and ethanol process yields as follows:

a_x = kmols of ATP required per kg of cells formed

a_{pl} = kmols of ATP formed per kg of ethanol produced
(hence a_x/a_{pl} is the amount of ethanol formed per kg of cells produced.

$Y_{x/s}$ = yield coefficient for cells on carbohydrate   = 0.82

$Y_{p/s}$ = yield coefficient for ethanol on carbohydrate = 0.51

For the yield of cells:

Formation of 1 kg of cells will utilise 1 / 0.82 kg of carbohydrate, using the energy released by the production of a_x/a_{pl} kg of ethanol. The amount of carbohydrate required for the ethanol will be (a_x/a_{pl}) / 0.51, so that the total amount of carbohydrate consumed to produce 1 kg of cells is

$$(1 / 0.82) + (a_x / 0.51\ a_{pl}).$$

The cell yield is therefore $\dfrac{1}{(1/\ 0.82) + (a_x/\ 0.51\ a_{pl})}$.

For typical values of a_x = 0.095 and a_{pl} = 1/ 46 the overall cell yield thus calculated is 0.102.

Similarly for ethanol:

Formation of 1 kg of ethanol will utilise 1/ 0.51 kg of carbohydrate and provide enough energy for a_{pl}/ax kg of biomass to be formed. This will consume (a_{pl}/a_x)/ 0.82 kg of carbohydrate and the overall ethanol yield is then, using the values given above:

$$\frac{1}{(1/\ 0.51) + (a_{pl}/\ 0.82\ a_x)} = 0.446$$

This figure for the ethanol yield is often stated as being 87% (0.446/ 0.51) of the theoretical (stoichiometric) yield. Note that we have not included any term for the maintenance energy requirement, which would further reduce the effective yields of both cells and ethanol.

Overall substrate utilization rates are expressed as the sum of the rates of substrate utilisation for the various products, including new cells as shown in 3.1. Where the energy for cell and product synthesis, as well as that for maintenance, is provided by oxidative phosphorylation, then CO_2 and water are amongst the products. Unfortunately these are also among the products of the synthesis reactions and it is not possible to distinguish how much arises from the different sources. However if the oxygen uptake rate is known, then the substrate utilisation for oxidative phosphorylation can be determined by the stoichiometry of the combustion reaction.

Thus for oxidation of glucose to CO_2 and water with molecular oxygen, we use 32 kg of oxygen for each 30 kg of glucose consumed. Therefore the yield coefficient for substrate required for oxygen, $Y_{o/s}$, is 1.067, and the rate of substrate uptake for this purpose is given by

$$r_{so} = r_o / Y_{o/s} \qquad (3.31)$$

Where the oxygen uptake rate for oxidative phosphorylation is not known, then the substrate needed to generate the energy required for synthesis is included with the substrate requirement for the synthesis itself according to the chemical equation. The two substrate requirements are lumped together and to distinguish the quantity $Y_{a/s}$ from a true stoichiometric coefficent we refer to it as a yield factor, since it includes both the elemental requirements and the energy requirements for the synthesis of component 'a'. The context usually determines which of these two descriptions applies and the symbol $Y_{a/s}$ is used rather indiscriminately for either. If it is required to distingusish between the two then the yield factor has the tick superscript $Y'_{a/s}$.

In those cases where we do not explicitly use the oxygen uptake rate, the substrate requirement to provide energy for maintenance is assumed to be first order with respect to cell concentration, i.e.

$$r_{sm} = m_s x_v \qquad (3.14)$$

Thus the more common expression for substrate utilisation (equation 3.1) is

$$r_s = r_{sm} + r_{sp1} + r_{sp2} + r_{sp3} + r_{sm} \qquad (3.32)$$

$$= \frac{r_x}{Y'_{x/s}} + \frac{r_{p1}}{Y'_{p1/s}} + \frac{r_{p2}}{Y'_{p2/s}} + \frac{r_{p3}}{Y'_{p3/s}} + m_s x_v$$

Any energy losses due to the slippage reaction ATP→ADP→AMP (with consequent additional heat release) will affect the values of the parameters $Y'_{a/s}$ and m_s.

3.7 Environmental Effects

3.7.1 Temperature effects

The individual reactions which occur within the cell are influenced by temperature in the usual way, i.e. the rate constant for the reaction is given by the Arrhenius equation in the form:

$$k = A \exp(-E/RT) \tag{3.33}$$

where E is the activation energy, T the absolute temperature and A is a constant. Since a large number of reactions influence the growth and product formation within the cell, the effect of temperature on the overall rates is complex. An additional complicating factor arises from the effects of temperature on the conformation of the enzyme proteins, in which very rapid changes in activity can take place over a small range of temperatures.

In an unstructured model, it is normal to express the parameters in the rate equations as Arrhenius functions of temperature; for example, the Monod equation for cell growth (above, 3.6) becomes:

$$r_x = \frac{\mu_m(T)\ S}{k_s(T) + S}\ x_v \tag{3.34}$$

indicating that both μ_m and k_s are functions of the temperature. By curve fitting on typical experimental data it is found that these functions usually have the forms:

$$\mu_m(T) = A_1 \exp(-E_1/RT) - A_2 \exp(-E_2/RT) \tag{3.35}$$

and $$k_s(T) = A_3 \exp(-E_3/RT) \tag{3.36}$$

In expression 3.35 for the maximum specific growth rate μ_m the first term on the right accounts for the general increase in reaction rate with temperature, whilst the second term (which normally has a much higher activation energy than the first) has been associated with a rapid reduction in the activity of the enzymes as temperature increases above a certain point.

The "optimum" temperature results from the interaction of these two effects. Its actual value varies greatly with the type of micro-organism, for while most common organisms have optimum temperatures in the "ambient" range, specialised organisms can have temperature optima at almost any temperature at which liquid water can exist. However for all organisms the relative behaviour at temperatures above and below the optimum is similar.

A typical plot of $\mu_m(T)$ versus T is shown in Figure 3.4. If an Arrhenius relation applies then a plot of ln(kinetic parameter) versus $1/T$ ($^{\circ}K^{-1}$) should either be a straight line or be made up from the sum or differences of two or more straight lines. Figure 3.5 shows these relationships.

Since the activation energies of the various reaction steps and conformational changes can differ widely it is to be expected that the optimum temperatures for growth and product formation may be quite different. Indeed in a well balanced medium the limiting substrate may also change with temperature. Control of reactor temperature to within 0.5° centigrade is usually required for efficient operation.

3.7.2 Effects of pH

The external pH in a fermentation usually has less of an effect on biological activities than does the temperature, because the cell is reasonably well able to regulate its internal hydrogen ion concentration in the face of adverse external concentrations, though the maintenance energy required is obviously affected. In addition the pH of the external medium probably has an important effect on the structure and permeability of the cell membrane.

A typical plot of the specific growth rate of a cell against the external pH is shown in Figure 3.6. There is a fairly wide range of pH over which the growth rate varies by little; this is often centred around pH 7 for most bacteria and pH 4.5 for fungi, but (as with temperature) there are important groups of microorganisms with very different pH optima. Where necessary, curves of growth rate versus pH can be modelled by expressions used in enzyme kinetics, of which one example might be:

$$\mu_m(pH) = \frac{\mu_m}{1 + (k_1/[H^+]) + k_2[H^+]} \qquad (3.37)$$

where $[H^+]$ [$=$ antilog $(-pH)$] is hydrogen ion concentration.

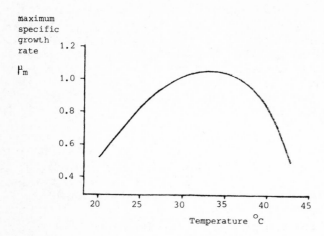

FIGURE 3.4. Effect of temperature on specific growth rate.

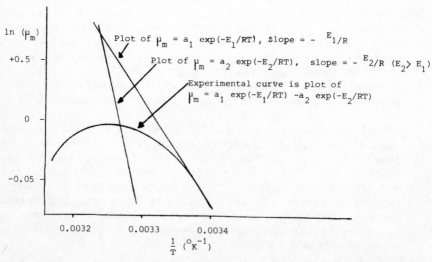

To decompose experimental curve draw straight line asymptote at low T (high 1/T). Subtract the experimental μ_m values (not $\ln(\mu_m)$ values) from μ_m values given by the asymptote. Plot the difference on the Arrhenius plot (i.e. ln(difference) vs 1/T). this should give the second straight line. If not, adjust the first asymptote. If it is not possible to get a straight line then repeat decomposition on curve.

FIGURE 3.5. Arrhenius plot of growth rate constant.

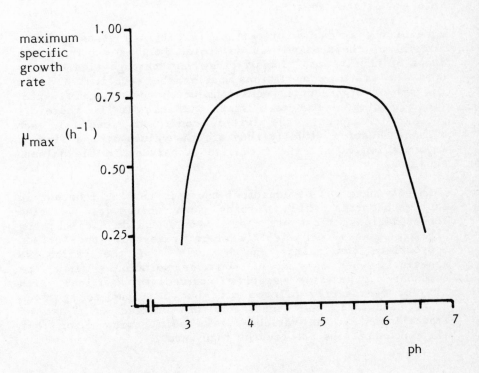

FIGURE 3.6. Effect of pH on specific growth rate.

4.1 Constraints and Initial Conditions

Each time we write a balance equation for any extensive property, we are also specifying the manner in which the extensive property will vary through time.

This is done by solving the differential equation which results from substitution of the rate equations (Chapter 2) and thermodynamic relationships (sections 1.2 and also 9.1) into the balance equation.

Each of the extensive properties is called a **state variable** and all of their values at any point in time determine the state of the system. The way they vary through time depends upon the starting conditions and on the equations i.e. on the set of differential equations which govern their variation in time. For each state variable it is therefore necesary to specify the **initial** conditions, one for each balance equation. Usually they are the values at the time of innoculation of all the extensive properties in the balance equations.

When solving a set of equations there is nothing inherent in the mathematics which forbids such things as negative concentrations, flows etc., even when these do not make any physical sense. It is therefore necessary to include appropriate **constraints** on the values of the variables, especially when relying on computer solutions, since the computer may generate negative intermediate solutions which are not necessarily printed out, but can lead to spurious final solutions. The constraints are written as a set of inequalities on the variables, the most common being that all concentrations are greater than zero.

4.2 Steady State and Unsteady State Models

The generalized method of model writing we have presented thus far makes no conditions as to the constancy of the variables through time. This is called an unsteady state model since the state of the system can vary. With most physical systems encountered in biotechnology, if the inputs to a system are fixed, the system will eventually come to a steady state where all the variables are constant; it may· of course be that all the cells are dead, but nevertheless the cell concentrations and the substrate concentration will be uniform!

One way of determining what this steady state is would be to solve the equations and allow time to proceed until every variable had reached its steady state value. Indeed this is a recognised method of solving for steady state values where the steady state equations are particularly difficult to solve. However it may be simpler to solve for the special case of steady state directly. Since all the variables are constant at the steady state, the rates of accumulation (i.e. the rate of change of the extensive variables) must all be zero. We can therefore reduce the unsteady state model to a steady state model by setting all the derivatives with respect to time to zero. The balance equations now become:

$$\text{Input - Output } = 0 \qquad\qquad (4.1)$$

The balance equations are now algebraic instead of differential, and in simple cases, some of which will be discussed later, they can be solved analytically. In more general cases both forms may be equally difficult to solve.

4.3 Checking the Model

It is not unusual for the first attempt at writing a mathematical model to result in a set of equations which do not have a solution which makes physical sense, or which are not solvable even by computer methods. Normally, this is because the problem was wrongly specified in the first place. Although this may be remedied by rechecking the physical and conceptual background of the model, a brief check of the mathematical consistency of the set of equations will often reveal some of the errors.

The model consists of a set of balance equations, one for each extensive property of the system considered to be among the variables of interest. As mentioned above, these are also called the state variables of the system. All the other variables in the equations must either be independent variables, i.e. those which are fixed externally to the system by an operator or some external control mechanism (e.g. the dilution rate, temperature, inlet substrate concentration etc.) or else can be expressed as functions of the state variables or the independent variables. As an example of the latter type, consider a kinetic rate expression such as

$$r_x = \frac{\mu_m \, S}{k_s + S} \, x_v \qquad (3.6)$$

Here r_x is a variable which is a function of the two state variables S and x_v.

The consistency check is to apply the equality
$$I = V - E \qquad (4.2)$$

where   I   is the number of independent variables which must
         be externally specified
        V   is the total number of variables
        E   is the number of equations we have written.

Having written the model and checked it for mathematical consistency, the next step is to check it for physical sense.

To do this the model is solved for a wide variety of conditions, analytically or by computer, and the physical sense of the solutions is verified. If the model is at all complex, considerable time will often be saved by checking that simplifying assumptions reduce it to models which are already well known and proven in the literature. A number of obvious checks are listed below.

1.   Check that the equations are dimensionally consistent.
2.   Reduce to the <u>steady state form</u> by setting all
      differentials to zero.
3.   Reduce to the <u>pure batch form</u> by setting all inflows
      and outflows to zero.
4.   Write all kinetic expressions as first order expressions
      and check the solutions.
5.   Check that the limit of the steady state solution as the
      dilution rate is reduced to zero is the limit of the
      pure batch solution as time approaches infinity.
6.   Check that the model behaves reasonably at
      understandable extremes of the independent variables.

Finally we should reiterate the golden rule: start simple and work up to the more complex. As stated in section 1.2,

<u>Always use the simplest adequate model firmly rooted in</u> .
<u>known fundamental physical, chemical and biochemical ideas</u>.

Preliminary note.

Before applying the general models developed in chapters 2 and 3 to specific fermentation processes, it will be useful as a reminder to list and classify the different variables and parameters.

a) State Variables

These define the state of the process and there is one for each extensive property:

x_v Viable cell concentration

x_d Non-viable cell concentration

S Outlet and fermenter limiting substrate concentration

P Outlet and fermenter product concentration

b) Operating Variables

These are variables whose values can be set by the operator of the process:

D Dilution rate (or F feed flowrate)

S_i ,x_i ,x_{di} ,P_i Inlet concentrations of the four conserved quantities. Often only S_i need be considered and x_i,x_{di} and P_i are equal to zero.

c) Intermediate variables

These are all the rates r_x, r_d, r_{sx}, r_{sm}, r_{sp} and r_p which can all be expressed in terms of the state variables listed above.

d) Kinetic parameters

As already defined: μ_m, k_s, k_d, m_s, α, β, etc.

e) Stoichiometric parameters

As already defined: $Y_{x/s}$, $Y_{p/s}$ etc.

5.1 Kinetic models for batch fermentations

The batch fermenter is the simplest and most common example of an unsteady state process. It is assumed that the reactor content is well mixed and that evaporation of water or other low boiling materials from the medium is negligible (or is constantly replaced). Hence the volume of medium within the fermenter is constant at the intial conditions. In Chapter 2 we presented the general form of the balance equation on a unit volume (of liquid) basis, equation 2.16, namely,

$$\frac{1}{V}\frac{d(Vy)}{dt} = \sum r_{gen} - \sum r_{cons} + Dy_i - \gamma Dy_o \qquad (2.16)$$

where y is the general extensive property.

The left hand side of this equation can be written as:

$$\frac{1}{V}\frac{d(Vy)}{dt} = \frac{1}{V}\cdot\left\{V\frac{dy}{dt} + y\frac{dV}{dt}\right\} \qquad (5.1)$$

Thus, in a classical batch reactor, where $F_i = F_o = 0$, (that is, $dV/dt = 0$), equation 2.14 becomes

$$\frac{dy}{dt} = \sum r_{gen} - \sum r_{cons} \qquad (5.2)$$

The equations for the state variables now become:

viable cells:
$$\frac{dx_v}{dt} = r_x - r_d \qquad (5.3)$$

non viable cells:
$$\frac{dx_d}{dt} = r_d \qquad (5.4)$$

substrate:
$$\frac{dS}{dt} = -(r_{sx} + r_{sm} + r_{sp}) \qquad (5.5)$$

product:
$$\frac{dP}{dt} = r_p$$

Having written the material balances the next step in producing a model is to choose a suitable set of kinetic rate equations. Starting with the simplest expressions (with the exception of the product formation in which case the starting point is invariably the general equation 3.19) from chapter 3 we have:

rate of cell growth:
$$r_x = \mu x_v \qquad (3.7)$$

rate of cell death:
$$r_d = k_d x_v \qquad (3.17)$$

rate of product formation:
$$r_p = \alpha r_x + \beta x_v \qquad (3.19)$$
$$= \alpha \mu x_v + \beta x_v$$

rate of substrate consumption for biomass production:
$$r_{sx} = \frac{r_x}{Y'_{x/s}} \qquad (3.29)$$

$$= \frac{\mu x_v}{Y'_{x/s}}$$

rate of substrate consumption for product formation:
$$r_{sp} = \frac{r_p}{Y'_{p/s}} \qquad (3.28)$$

$$= \frac{\alpha \mu x_v + \beta x_v}{Y'_{p/s}}$$

rate of substrate consumption for maintenance energy:
$$r_{sm} = m_s x_v \qquad (3.28)$$

Substituting these relationships into the material balances we get:

viable cells:
$$\frac{dx_v}{dt} = \mu x_v - k_d x_v \qquad (5.7)$$

non-viable cells:
$$\frac{dx_d}{dt} = k_d x_v \qquad (5.8)$$

substrate:
$$\frac{dS}{dt} = - \left\{ \frac{\mu x_v}{Y'_{x/s}} + m_s x_v + \frac{\alpha \mu x_v + \beta x_v}{Y'_{p/s}} \right\} \qquad (5.9)$$

product:
$$\frac{dP}{dt} = \alpha \mu x_v + \beta x_v \qquad (5.10)$$

As in section 2.4, if the concept of endogeneous respiration is incorporated into the model rather than that of maintenance energy, the model equations will become:

viable cells: $$\frac{dx_v}{dt} = \mu x_v - k_d x_v - k_e x_v \qquad (5.11)$$

non-viable cells: $$\frac{dx_d}{dt} = k_d x_v \qquad (5.12)$$

substrate: $$\frac{dS}{dt} = -\left\{ \frac{\mu x_v}{Y'_{x/s}} + \frac{\alpha \mu x_v + \beta x_v}{Y'_{p/s}} \right\} \qquad (5.13)$$

product: $$\frac{dP}{dt} = \alpha \mu x_v + \beta x_v \qquad (5.14)$$

To complete the process model for the batch fermentation we must also include a set of initial conditions, which will be the values of all the state variables (concentrations) at the time of inoculation, together with the constraints as outlined in chapter 4.

An analytical solution of this set of equations is possible but is too complex to be generally useful. Computer solution by numerical integration is more usual, and a typical solution is illustrated in Figure 5.1.

Two features that this plot does not show, but which are actually present in most fermentations, are the "lag" phase and the phase of decline.

The lag phase arises from the need for cells in the inoculum to adapt to the prevailing conditions in the fermenter, and the model as constructed does not incorporate any mechanism for this adaptation (structured models are required to do this, see 2.3.2).

The phase of decline would only be shown if we had included cell lysis or endogeneous respiration in the model; in chapter 2 we assumed that the non-viable cells remain unchanged in the culture so that the total mass of viable and non-viable cells remains constant after the limiting substrate has been exhausted.

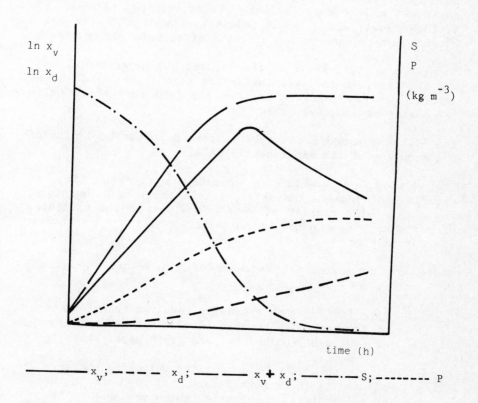

FIGURE 5.1. Biomass and product formation in batch culture
derived by computer simulation of equations
5.7–5.10.

5.2 Measuring and Quantifying Kinetic Parameters

Having constructed a model, the next task is to determine actual values for the model parameters. These parameters can be subdivided into stoichiometric (those concerned with the material and energy balance relationships between the various flows) and kinetic (those concerned with the rates of consumption, generation or rates of transfer of species).

To determine the values of the parameters, experimental data are required. It goes without saying that these should be as accurate and as reliable as possible, and they should also be consistent and replicable.

The 'fit' of a model can only be tested up to the limits of the accuracy of the experimental data.

The method of quantifying parameters is to fit the experimental data to the appropriate mathematical equations. These data will consist of sets of values of the variables of interest determined simultaneously at a particular instant in time.

There are two principal variants of the method of fitting data to the model:

<u>Straight line fitting</u> - this usually involves transformation of the variables but can be done with the minimum of equipment (calculator and graph paper etc.)

<u>Curve fitting</u> - this is usually carried out using the sets of data in their normal form, but requires the use of a computer and software, which may need to be tailored for the particular class of model being fitted.

The most familiar example of straight line fitting is the use of the Lineweaver-Burke plot to determine the maximum rate and saturation constant of a Michaelis-Menten or Monod type rate equation, such as equations 3.5 and 3.6 which are both of this form. Taking equation 3.6 as our example, we have:

$$r_x \quad = \quad \frac{\mu_m S}{k_s + S} \cdot x_v \qquad\qquad (3.6)$$

where r_x, S and x_v are the variables and μ_m and k_s are the parameters. As it stands the relationship betwen the variables is non-linear (i.e. involves multiplication or division). The first step is to transform the variables i.e. to manipulate them in some way so as to give a linear relationship between the new transformed variables. This procedure is not formalised and depends on the ingenuity of the modeller. In this case division of r_x by x_v simplifies the expression by removing a multiplication. Thus by defining the new variable $\mu = r_x/x_v$ (μ is the specific growth rate) we have transformed the equation to:

$$\mu = \frac{\mu_m \, S}{k_s + S} \qquad (5.15)$$

where now we have just two variables μ and S. The next step is to take the reciprocal of either side of the equation to give:

$$\frac{1}{\mu} = \frac{k_s + S}{\mu_m \, S} = \frac{k_s}{\mu_m} \frac{1}{S} + \frac{1}{\mu_m} \qquad (5.16)$$

or in terms of new transformed variables:

$$p = \frac{k_s}{\mu_m} + \frac{1}{\mu_m} \qquad (5.17)$$

where $\qquad p = x_v/r_s$ and $q = 1/S$

This equation is now linear in the transformed variables, p and q, so that a plot of p versus q (on ordinary, i.e. arithmetic, graph paper) should give a straight line with a slope that is equal to k_s/μ_m and an intercept that is $1/\mu_m$. A typical plot of this kind is shown here as Figure 5.2. Geometrical consideration show that the intercept on the 1/S axis is equal to $-1/k_s$. Thus from the two intercepts, μ_m and k_s are determined.

This type of plot is rather generally useful, and because the experimenters carry out the transformation and plotting themselves, they should be able to take account of the relative accuracy of the data and the distortions introduced by the transformations. Thus for example low values of S are usually less accurate in a relative sense than are high values. Consequently the high values of 1/S (corresponding to low S) will be relatively inaccurate. This affects the estimate of k_s (from the intercept) more than that of μ_m, and emphasizes the experimental necessity of obtaining as high a relative accuracy as possible (by duplication etc.) at low substrate concentrations.

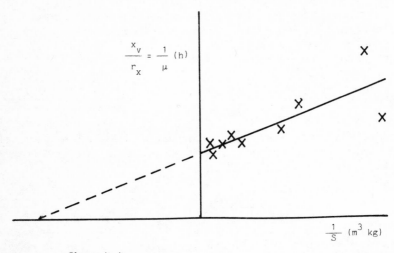

$$\frac{x_v}{r_x} = \frac{1}{\mu} \ (h)$$

$$\frac{1}{S} \ (m^3 \ kg)$$

Slope= k_s/μ_{max}
Intercept on $1/\mu$ axis= $1/\mu_{max}$
Intercept on $1/S$ axis= $-1/k_s$

FIGURE 5.2. Determination of kinetic parameters using a Lineweaver-Burk type plot. Note that the data is normally less accurate (more scatter) at low S (high 1/S).

Another example of the transformation of equations is the determination of the kinetic parameters in the equation for product formation viz:

$$r_p = \alpha r_x + \beta x_v \qquad (3.19)$$

To obtain values for the product formation parameters α and β, equation 3.19 is transformed by dividing by x , to give:

$$\frac{r_p}{x_v} = \alpha \frac{r_x}{x_v} + \beta x_v \qquad (5.18)$$

which is tested by plotting r_p/x_v against r_x/x_v , as shown in Figure 5.3.

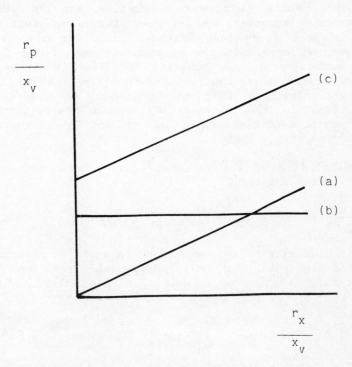

FIGURE 5.3. Determination of product formation kinetics according to the Luedeking–Piret model.

(a) growth associated $\alpha > 0$, $\beta = 0$

(b) non-growth associated $\alpha = 0$, $\beta > 0$

(c) intermediate kinetics α , $\beta \geq 0$

6.1 The Chemostat

The classical chemostat is the system to which most "continuous culture" data in the literature relate, or are intended to conform; it is a powerful research tool since it allows us to obtain steady-state data for the interactions between micro-organisms and their environment, and industrially it is the most efficient system for certain practical purposes, though by no means for all.

A chemostat is a fermenter which operates with a constant inflow and outflow, and is assumed to be so well mixed that all the concentrations are uniform throughout the whole of the medium volume. **Outflow concentrations are the same as those in the fermenter,** and constant inflow and outflow of medium ensures a constant reactor volume. In the general equations, V represents the constant volume of the liquid phase, excluding any gas bubbles. The final requirement is that we assume that conditions have been uniform for a sufficient time, so that any transient changes due to fluctuations in the flow rates, medium compositions or environmental conditions have died out and the fermenter is operating at steady state.

The general form of the balance equation on a unit volume basis is, from chapter 2:

$$\frac{1}{V}\frac{d(Vy_o)}{dt} = \sum r_{gen} - \sum r_{cons} + D(y_i - \gamma y_o) \qquad (2.16)$$

The left hand side of this equation can again be expanded as explained in the previous Chapter (see equation 5.1). However for the chemostat, since $F_i = F_o$, $dV/dt = 0$ and $\gamma = F_o/F_i = 1$, so that equation 2.16 now becomes:

$$\frac{dy_o}{dt} = \sum r_{gen} - \sum r_{cons} + D(y_i - \gamma y_o) \qquad (6.1)$$

where $D = F/V$ is the dilution rate (h^{-1}). This is a key parameter, whose reciprocal is the mean residence time of all the materials flowing through the system.

In a chemostat operating at steady state, there is no accumulation of any extensive quantity, thus the specific material balance in every case can be obtained from the general equation 6.1 by setting $dy_o/dt = 0$.

Thus we will have, for-

viable cells: $\quad\quad 0 = r_x - r_d + D(x_{vi} - x_{vo})$ $\quad$ (6.2)

non-viable cells: $\quad 0 = r_d \quad\quad\quad + D(x_{di} - x_{do})$ $\quad$ (6.3)

substrate: $\quad\quad 0 = -(r_{sx}+r_{sm}+r_{sp}) + D(S_i - S_o)$ $\quad$ (6.4)

product: $\quad\quad\quad 0 = r_p \quad\quad\quad + D(P_i - P_o)$ $\quad$ (6.5)

Note again the symmetry of the equations, and also that the equation for total volume now disappears completely since the volume does not change.

This above set of equations, together with the kinetic rate equations and the constraints, represent the general mathematical model of a chemostat.

To determine the relationships between the state variables (x_{vo}, x_{do}, S_o and P_o) and the operating variable 'D, equations 6.2 to 6.5 can all be solved analytically with x_{vi}, x_{di} and P_i all zero. This is the most common case, when the input is free from cells and contains no cell products.

The method can be illustrated by modelling a simple fermentation process in which substrate is only converted into viable cells and there is no metabolite production. As a first approximation for simplicity we will also ignore cell death and maintenance, so that we can put

$$r_d = r_{sm} = 0.$$

The equation for the chemostat now becomes

$$0 = r_x - Dx_{vo} \quad\quad\quad\quad\quad\quad (6.6)$$

$$0 = -r_{sx} + D(S_i - S_o) \quad\quad\quad\quad (6.7)$$

Assuming the cells grow according to Monod kinetics, we can also use equation 3.7 and 3.29, obtaining:

$$0 = \mu x_{vo} - Dx_{vo} \quad\quad\quad\quad\quad\quad (6.8)$$

$$0 = -\frac{\mu x_{vo}}{Y'_{x/s}} + D(S_i - S_o) \quad\quad\quad (6.9)$$

From equation 6.8 we get

$$\mu = D \qquad (6.10)$$

and this can be substituted into equation 6.9 to give

$$x_{vo} = Y'_{x/s} (S_i - S_o) \qquad (6.11)$$

The substrate concentration S_o is obtained from the Monod expression:

$$\mu = \mu_m \frac{S_o}{k_s + S_o} \qquad (3.8)$$

or by rearranging and substituting equation 6.10

$$S_o = \frac{k_s D}{\mu_m - D} \qquad (6.12)$$

These equations should all be familiar to the reader, even if this explanation of their origin is new. Note that what we have done is to set up first the more general case (equations 6.2 - 6.5), and then derive the more particular case from it. This should give us a clearer understanding of the special assumptions that have been made. A typical solution for this case is illustrated here in Figure 6.1.

To solve the general model equations 6.2 to 6.5 without these simplifying assumptions, we must select suitable kinetic rate expressions. As before (ch. 3) we shall use

$$r_x = \mu x_v \qquad (3.7)$$

$$r_d = k_d x_v \qquad (3.17)$$

$$\begin{aligned} r_p &= \alpha r_x + \beta x_v \\ &= (\alpha\mu + \beta)x_v \end{aligned} \qquad (3.19)$$

$$r_{sx} = \frac{\mu x_v}{Y'_{x/s}} \qquad (3.29)$$

$$r_{sp} = \frac{r_p}{Y'_{p/s}} = \frac{(\alpha\mu + \beta)x_v}{Y'_{p/s}} \qquad (3.28)$$

$$r_{sm} = m_s v_v \qquad (3.14)$$

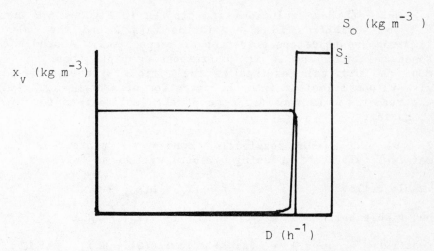

FIGURE 6.1. Steady state biomass and limiting substrate
concentration versus dilution rate according to
Monod kinetics.

Using these expressions, the model solution for the general
chemostat case can be obtained using the procedure outlined
above. The reader should verify for him or herself that the
resulting set of equations, set out below, is mathematically consistent, as defined in section 4.3.

$$x_{vo} = \frac{D(S_i - S_o)}{\dfrac{(D + k_d)}{Y'_{x/s}} + m_s + \dfrac{\alpha(D + k_d) + \beta}{Y'_{p/s}}} \qquad (6.13)$$

$$x_{do} = \frac{k_d x_v}{D} \qquad (6.14)$$

$$S_o = \frac{k_s(D + k_d)}{\mu_m - (D + k_d)} \qquad (6.15)$$

$$P_o = \frac{[\alpha(D + k_d) + \beta] x_v}{D} \qquad (6.16)$$

Constraints:

$$0 \leqslant S_o \leqslant S_i \; ; \; 0 \leqslant x_{vo} \; ; \; 0 \leqslant x_{do} \leqslant x_{vo} \; ; \; 0 \leqslant P_o$$

A sample of these solutions are plotted in Figures 6.2 and 6.3 for variations in the dilution rate D and the inlet limiting substrate concentration S_i respectively. A study of these solutions and of the corresponding plots, bearing in mind the physical meanings of the various symbols, should give valuable insight into the operation of a chemostat, and the reader should consider some of the implications for him or herself.

If the endogeneous respiration concept is preferred, the material balances (replacing 6.2-6.5) will be as follows:

viable cells: $$0 = r_x - r_d - r_e + D(x_{vi} - x_{vo}) \qquad (6.17)$$

non-viable cells: $$0 = r_d \qquad\qquad + D(x_{di} - x_{do}) \qquad (6.18)$$

substrate: $$0 = -(r_{sx} + r_{sp}) + D(S_i - S_o) \qquad (6.19)$$

product: $$0 = r_p \qquad\qquad + D(P_i - P_o) \qquad (6.20)$$

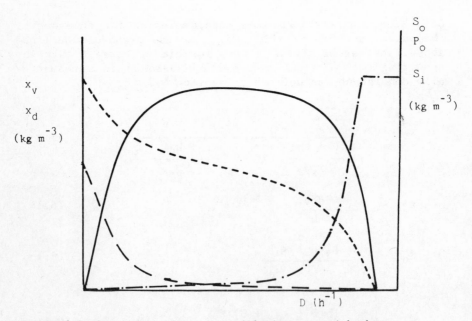

FIGURE 6.2. State variables versus dilution rate with the concentration of inlet limiting substrate S_i constant. (N.B. all other substrates including oxygen are present in stoichiometic excess).

——————— x_v; ——— — x_d;—·—·—— S_o; ------- P_o

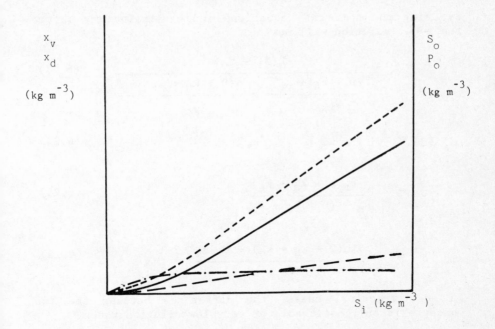

FIGURE 6.3. State variables versus inlet concentration of
limiting dissolved substrate at constant dilution
rate.   (N.B. all other substrates including
oxygen must be present in stoichiometic excess).

———————— x_v ; ———— x_d ; —·—·— S_o ; ——————— P_o

If we use the following rate expressions,

$$r_x = \mu x_v ; \qquad r_d = k_d x_v ; \qquad r_e = k_e x_v ,$$

then equation 6.17 becomes,

$$0 = \mu x_{vo} - k_d x_{vo} - k_e x_{vo} + D(x_{vi} - x_{vo}) \qquad (6.21)$$

When $x_{vi} = 0$, as is normally the case, equation 6.21 will
give the following expression for the specific growth rate:

$$\mu = D + k_d + k_e \qquad (6.22)$$

This should be compared with equation 6.10.

by using the same rate equations and constraints as before the model solution will now be:

$$x_{vo} = \frac{D(S_i - S_o)}{\dfrac{(D + k_d + k_e)}{Y'_{x/s}} + \dfrac{\alpha(D + k_d + k_e) + \beta}{Y'_{p/s}}} \tag{6.23}$$

$$x_{do} = \frac{k_d x_v}{D} \tag{6.24}$$

$$S_o = \frac{k_s(D + k_d + k_e)}{\mu_m - (D + k_d + k_e)} \tag{6.25}$$

$$P_o = \frac{[\alpha(D + k_d + k_e) + \beta]\, x_v}{D} \tag{6.26}$$

In most practical cases, the difference between the two concepts is only noticeable at very low dilution rates.

An analytical solution is not always obtained so readily, and then we have to resort to computer solutions and be content with graphical plots.

It is because computer solutions can generate utter rubbish if not handled carefully, that careful specifications of the constraints on the variables are a necessary part of the model.

One aim in using models of this kind is to allow prediction of the behaviour of the chemostat when conditions are altered, whether by changing the operating variables or by the influence of changed environmental conditions on the values of the parameters. This of course requires that the effect of changes in the environmental conditions on the parameters is already known, but conversely the chemostat is ideally suited for determining these effects, and this is one of its main research applications. However use of the chemostat in this way requires a satisfactory model if the results are to be properly understood.

Chapter 8 illustrates how models may be used for this purpose.

6.2 Measuring and quantifying kinetic parameters

The principles involved in measuring and quantifying kinetic parameters for batch systems have been described in section 5.2. In continuous culture, the experimental data which will be fitted to the appropriate mathematical equations will consist of sets of values of the variables of interest, determined either when the system has settled down at steady state, or determined simultaneously at some given instant in time as in batch culture (as in observing unsteady states, in dynamic experiments). As in considering batch culture in section 5.2, we will concentrate on straight line fitting, invariably the method of choice when steady state data are to be fitted to the relatively simple models included in this book.

Again as for batch culture, the parameters μ_m and k_s are determined from a Lineweaver-Burke plot (see figure 5.1). Using the same equations as before (equations 3.6 and 5.15) we introduce equation 6.10 to get a steady state chemostat version of the Lineweaver-Burke plot. By substituting the chemostat equality $\mu = D$, equation 5.16 becomes:

$$\frac{1}{D} = \frac{k_s}{\mu_m} \cdot \frac{1}{S} + \frac{1}{\mu_m} \tag{6.27}$$

and the parameters can be determined from a plot of $1/D$ against $1/S$.

In this case we already had a model which could be transformed to a linear relationship from which two parameters can be determined. Normally, however, the number of kinetic parameters in any one expression is greater than two and this direct procedure must be extended. A typical example would be the steady state solution to the product formation equation:

$$P_o = \frac{[\alpha(k_d + D) + \beta] \, x_{vo}}{D} \tag{6.16}$$

By dividing by x_{vo} this equation can be transformed to:

$$\frac{P_o}{x_{vo}} = \alpha(k_d + \beta) \cdot \frac{1}{D} + \alpha \tag{6.28}$$

from which it is possible to obtain any two of α, β or k_d by plotting P/x_v against $1/D$.

Note how this last example allows for the estimation of two parameters only if the third is known. Thus we can estimate α from the intercept and either β or k_d from the slope, given that the other is known. For example, we may already be satisfied that k_d can usefully be put $= 0$.

Equation 6.15 provides another example where there are three parameters, and thus by analogy with the above case we would expect that we could only estimate two of the parameters given that the third was known. However this is an example of an equation where one of the parameters (k_d in this case) occurs only in a linear association with one of the variables (D).

We can therefore use the observation that there will be only one value of this parameter for which we will be able to obtain a straight line plot. The transformed equation is:

$$\frac{1}{S_o} = \frac{\mu_m}{k_s} \cdot \frac{1}{(D + k_d)} + \frac{1}{k_s} \qquad (6.29)$$

which will give a Lineweaver-Burke type of plot. Thus the procedure is to plot $1/S_o$ against $1/(D + k_d)$ for various values of the parameter k_d. For only one of these k_d values will we get a straight line and all others will give a curve. Hence this one value gives us the value of k_d which best fits the data, and subsequently μ_m and k_s can be calculated.

Figure 6.4 illustrates this technique.

Where the number of parameters to be fitted using one equation or set of equations is too large to adopt these straight line fitting techniques, then some form of curve fitting must be used.

These techniques are beyond the scope of this chapter, and the appropriate mathematical or computer texts should be consulted. Whatever program is used it is essential that the experimenter should understand the basic principles of its operation before using it, and should always make sure that the values of the parameters generated make sense. As many of the parameters as possible should have been determined, singly or in pairs, by authenticated methods in experiments designed solely to elucidate these values, either by the experimenter himself or by others as published in the literature.

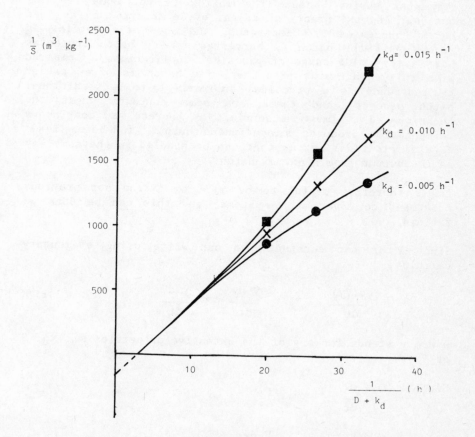

FIGURE 6.4. Selection of parameter value (k_d) to give straight
line for determination of parameters in
Lineweaver-Burk plot. The slope of the straight
line is u_{max}/k_s and the intercept on the 1/S axis
is equal to $-1/k_s$.

7.1 The Effect of Variable Volume

In a variable volume process, the volume of medium in the fermenter varies because the inflow is not equal to the outflow; in most practical cases, often referred to as "fed batch" processes, V increases, and often the outflow is zero (until the batch is harvested). Fed batch processes have a very wide range of practical applications. Compared with fully-continuous processes, fed batch is often easier to introduce as a practical improvement to a traditional batch process, and there are some contexts where the technique is almost essential. Whereas a continuous fermentation produces a continuous output, fed-batch leads to a single final output that can be handled in the same way as the output from a normal batch.

It is therefore useful to be able to extend our previous treatment to cover such systems, and this can be done as follows.

Since V is not constant, we can write using elementary calculus,

$$\frac{d(Vy)}{dt} = \frac{V \, dy}{dt} + \frac{y \, dV}{dt} \qquad (7.1)$$

where y stands for any of the extensive properties x_v, x_d, S or P.

Since

$$\frac{dV}{dt} = F_i - F_o \qquad (7.2)$$

then

$$\frac{d(Vy)}{dt} = \frac{V \, dy}{dt} + y(F_i - F_o) \qquad (7.3)$$

and

$$\frac{1}{V} \frac{d(Vy)}{dt} = \frac{dy}{dt} + Dy(1 - \gamma) \qquad (7.4)$$

for any extensive property y, where, as before,

$$D = F_i/V \quad \text{and} \quad \gamma = F_i/F_o$$

Combining equations 2.14 and 7.4, we can write the general balance equation as:

$$\frac{dy}{dt} + Dy(1 - \gamma) = \sum r_{gen} - \sum r_{cons} + D(y_i - \gamma y_o) \qquad (7.5)$$

Since we presume $y = y_o$ this gives the somewhat surprising final result:

$$\frac{dy}{dt} = \sum r_{gen} - \sum r_{cons} + D(y_i - y_o) \qquad (7.6)$$

This is identical in appearance to the general form already given for the balance equation 2.14 as applied to a constant volume process ($\gamma = 1$) given in section 2.4 !

However although the term $D(y_i - y_o)$ is identical in the two equations *it does not have the same physical significance.* In both the constant volume and the variable volume process, the term Dy_i represents the amount of the extensive quantity brought in with the inflow. In the constant volume process the term Dy_o is correspondingly the amount of the extensive quantity leaving with the outflow. However in the variable volume process, the term Dy_o is a combination of:

the amount leaving in the outflow $= \dfrac{F_o}{V} y_o = \dfrac{\gamma F_i}{V} y_o = \gamma D y_o$

and the volume dilution effect, which

$$= \frac{F_i - F_o}{V} y_o = \frac{F_i[1 - (F_o/F_i)]}{V} y_o = D(1 - \gamma) y_o$$

Thus: $\dfrac{F_o}{V} y_o + \dfrac{F_i - F_o}{V} y_o = \gamma D y_o + D(1 - \gamma) y_o = D y_o$

so that the outflow effect cancels out part of the dilution effect.

When there is no outflow from the fed batch the dilution effect is due solely to the inflow ($D = F_i/V$). Note that the values of y are still specified as y_o since we have used this subscript to emphasize the general case that values of variables in the outflow are the same as those prevailing in the control volume.

The full set of balance equations for the variable volume process in the general case is as follows:

Viable cells $\quad \dfrac{dx_{vo}}{dt} = r_x - r_d + D(x_{vi} - x_{vo}) \quad$ (7.7)

Non-viable cells $\dfrac{dx_{do}}{dt} = r_d \quad\quad + D(x_{di} - x_{do}) \quad$ (7.8)

Substrate $\quad \dfrac{dS_o}{dt} = -(r_{sx}+r_{sm}+r_{sp}) + D(S_i - S_o) \quad$ (7.9)

Product $\quad \dfrac{dP_o}{dt} = r_p \quad\quad + D(P_i - P_o) \quad$ (7.10)

Volume $\quad \dfrac{dV}{dt} = F_i - F_o \quad$ (7.11)

Note that $D = F_i/V$

These equations together with the set of kinetic rate equations, the initial conditions and the constraints form the mathematical model of the fed batch process.

7.2 Modelling Examples of Fed Batch

If we assume that $x_{vi} = x_{di} = P_i = F_o = 0$, we have the classical fed batch process, typically for type 2 and type 3 product formation, with a continuous input of the limiting substrate, usually a carbon energy source, in a feed which is free from cells and with no outflow from the vessel until the batch run is completed. As a possible set of kinetic rate equations we could take:

for growth $\quad\quad r_x = \mu x_v = \dfrac{\mu_m S}{k_s + S} x_v \quad$ (3.6)

for death $\quad\quad r_d = k_d x_v \quad$ (3.17)

for substrate uptake, $\quad r_{sx} = r_x / Y'_{x/s} \quad$ (3.29)

$\quad\quad\quad r_{sm} = m_s x_v \quad$ (3.14)

$\quad\quad\quad r_s = r_p / Y'_{p/s} \quad$ (3.28)

and for product formation, $r_p = \beta x_v \quad$ (3.20)

Note that in the normal process situation the operator can manipulate both S_i and F_i, and as will be seen later, from a purely mathematical point of view it is better to have high values of S_i and low values of F_i (i.e. a concentrated input feed). However considerable practical difficulties in achieving a uniform (and low) concentration of sugar or other limiting substrate in the fermenter may arise because of mixing limitations, and so there is a practical limit on the maximum useful value of S_i.

This particular model is an example of the use of mathematical models for "what if " type of computer experiments, as there will be a variety of objectives in running a fed batch project which can be very easily explored with simple extensions of the model to take into account some of the equipment factors.

One possible objective in a typical practical application of fed batch methods might be to have zero growth and high product formation rates.

The high product formation rate depends on having a high viable cell concentration, as can be seen from the product balance equations above (7.10 and 3.20 above). The first stage of the process would therefore be to grow up a high cell concentration, followed by a phase where growth is supressed and only sufficient of the substrate is supplied for maintenance and product formation.

We can consider this second stage separately, and will have:

for zero growth , $\mu = 0$, hence:

$$S = 0, \quad \frac{dS}{dt} = 0 \quad \text{and} \quad r_x = 0$$

Combining the kinetic rate equations with the balance equations then gives:

$$\frac{dx_v}{dt} = -k_d x_v - D\, x_v \qquad (7.11)$$

$$\frac{dx_d}{dt} = k_d x_v - D\, x_d \qquad (7.12)$$

$$0 \quad = \quad - \left\{ m_s x_v + \frac{x_v}{Y'_{p/s}} \right\} + D \, S_i \qquad (7.13)$$

$$\frac{dP}{dt} \quad = \quad \beta x_v \, - \quad D \, P \qquad (7.14)$$

$$\frac{dV}{dt} \quad = \quad F_i \qquad (7.15)$$

Note that, as already indicated for this case, the terms involving D in the above equations represent simply the dilution effect of the incoming feed, and must not be confused with dilution rate effects in a true chemostat. The solution of these equations to give us an optimum operating system is simplest if we fix S_i and allow F_i to vary, as shown below.

The substrate balance (7.13) can be rearranged to give:

$$0 \quad = \quad -k \, S_i x_v \, + \, D \, S_i \qquad (7.16)$$

where
$$k \quad = \quad - \left\{ m_s x_v + \frac{x_v}{Y'_{p/s}} \right\} \frac{1}{S_i} \qquad (7.17)$$

which is a constant providing S_i is kept constant.

From 7.16 we get $D \; = \; kx_v$, which on substituting into 7.11 gives:

$$dx_v/dt \quad = \quad - \, (k_d \, + \, kx_v) \, x_v$$

This can be integrated (using a standard table of integrals) to give:

$$x_v \quad = \quad \frac{k_1 k_d \, \exp(-k_d t)}{1 \, - \, k_1 k \, \exp(-k_d t)} \qquad (7.18)$$

in which $k_1 \quad = \quad \dfrac{x_v(0)}{k_d \, + \, kx_v(0)} \qquad (7.19)$

$x_v(0)$ is the value of the viable cell concentration when the batch feeding phase begins

t is time from beginning the feeding phase.

The volume is obtained by substituting for F_i ($= kx_vV$) into equation 7.15, which since x_v has already been given as a function of time (equation 7.18) can be directly integrated.

The expressions for x_d and P are more complicated, but if we write

$$f(t) = \int_o^t x_v \, dt \qquad (7.20)$$

then the equation for P is

$$P = \frac{\beta}{k} \{1 - \exp[-k.f(t)]\} \qquad (7.21)$$

Typical plots of these expressions are given in Figure 7.1.

In the idealised case where there is no death of cells, i.e. $k_d = 0$ and the second equation in the model is eliminated, the analytical solutions are much simpler and are:

$$x_v = x_v(0) / [1 + kx_v(0)t] \qquad (7.22)$$

$$F = kx_v(0) V(0) \qquad (7.23)$$

$$P = \beta x_v(0)t / [1 + kx_v(0)t] \qquad (7.24)$$

As before $x_v(0)$ is the cell concentration at the beginning of the fed batch phase and $V(0)$ is the corresponding volume.

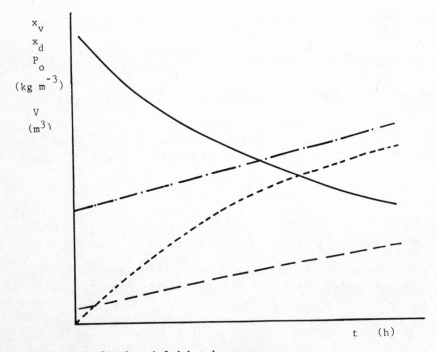

FIGURE 7.1. Simulated fed batch curves.

——————— x_v; ————— x_d; ·········· P_o; —·—·—· V

8.1 Washout in Continuous Culture

One consequence of the equations describing the simple chemostat is that the maximum dilution rate that can be used is limited to a value which must be slightly less than the maximum specific growth rate μ_m. This is called the critical dilution rate, D_c.

"Washout" occurs when cells are being removed from the fermenter at a rate $(D_c x_v)$ which is just equal to the maximum rate at which they can grow in the fermenter, which is $(\mu x_v - k_d x_v)$, so that the slightest increase in D would cause the steady state cell concentration to fall to zero. This maximum rate of growth in the fermenter occurs when

$$\mu = \mu_m \, S \, /(k_s + S_i),$$

since S_i is the maximum substrate concentration that can occur in the fermenter. As the value of k_s is small compared to S_i it follows that:

$$D_c = \mu_m \frac{S_i}{k_s + S_i} - k_d \triangleq \mu_m - k_d \qquad (8.1)$$

Of course at the actual point of washout the "steady state" is one in which no cells exist or grow and the substrate passes unchanged through the fermenter. In practice, mixing in the fermenter may be less than perfect and a few cells may remain behind. A sudden increase in S in the effluent, as the dilution rate is increased towards D_c, is usually taken as indicating the approach to washout. The rapidity of this effect depends on the relative magnitudes of k_s and S.

This limitation on the maximum dilution rate that can be used limits the productivity of the fermenter (Dx_v or DP, units kg.m^{-3} h^{-1}), which is maximal at a dilution rate somewhat less than D_c; the use of the technique known as cell recycle overcomes this limitation to some extent.

8.2 Recycle Systems

In a chemostat with cell recycle, all or part of the out-flowing medium is removed from the system either completely free of cells or with a cell concentration significantly less than that which exists in the chemostat. This situation may be achieved either within the chemostat, by filtering or settling part of the outflowing stream, or externally by using a separating device such as a filter, settler or a continuous centrifuge; different ways of doing this are shown schematically in Figure 8.1.

A working example of this kind of system would be the well-established "activated sludge" process for effluent treatment, which usually involves suitable combinations of arrangements (a) and (c) in figure 8.1.   More recent applications to such varied processes as ethanol ferment- ations and high-value bacterial product formation are generally still at the research stage,

However for the purposes of mathematical analysis they are all equivalent to having an outflowing stream with a cell concentration equal to the internal cell concentration multiplied by a constant, δ, which we will call the separation constant, $0 < \delta < 1$. An abstracted physical model of such a system, with nomenclature, is shown in Figure 8.2.

For all these systems, a material balance for the general extensive property, y, will again give equation 6.1, except that we write dy/dt on the left side, and $y = y_0$ only for substances which are dissolved in the liquid phase. In this case x_{vo} becomes δx_v and x_{do} becomes δx_d, where x_v and x_d refer to internal cell concentrations.

If the separation constant is zero this means that all the cells are retained in (or recycled back to) the fermenter and without cell lysis or endogeneous respiration the biomass in the chemostat would increase without limit; in practice gas transfer limitation or product inhibition would terminate this increase.
If δ equals unity then we have the simple chemostat without recycle.

The steady state equations are, modifying those given in chapter 6 for zero values of x_{vi}, x_{di} and P_i :

Viable cells $0 = r_x - r_d - \delta D x_v$ (8.2)

Non-viable cells $0 = r_d - \delta D x_d$ (8.3)

Substrate $0 = - (r_{sx}+r_{sm}+r_{sp}) + D(S_i - S_o)$ (8.4)

Product $0 = r_p - DP$ (8.5)

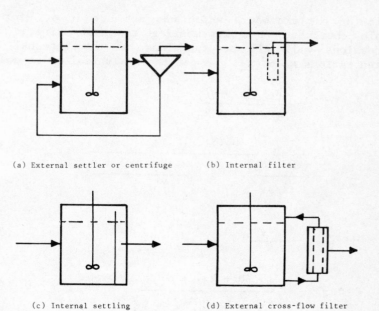

(a) External settler or centrifuge   (b) Internal filter

(c) Internal settling   (d) External cross-flow filter

FIGURE 8.1. Methods of recycling cells.

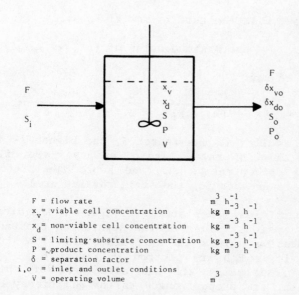

F = flow rate $m^3\ h^{-1}$
x_v = viable cell concentration $kg\ m^{-3}\ h^{-1}$
x_d = non-viable cell concentration $kg\ m^{-3}\ h^{-1}$
S = limiting substrate concentration $kg\ m^{-3}\ h^{-1}$
P = product concentration $kg\ m^{-3}\ h^{-1}$
δ = separation factor
i,o = inlet and outlet conditions
V = operating volume m^3

FIGURE 8.2. Chemostat with cell recycle.

Following the procedure which was set out in 6.1 for the simple chemostat, and remembering that the kinetic rate expressions will all be functions of the internal cell concentrations x_v or x_d, we get the analytical solutions:

$$S = \frac{k_s(\delta D + k_d)}{\mu_m - (\delta D + k_d)} \tag{8.6}$$

$$x_v = \frac{D(S_i - S_o)}{\dfrac{(\delta D + k_d)}{Y'_{x/s}} + m_s + \dfrac{\alpha(\delta D + k_d) + \beta}{Y'_{p/s}}} \tag{8.7}$$

$$x_d = \frac{k_d x_v}{\delta D} \tag{8.8}$$

$$P_o = \frac{[\alpha(\delta D + k_d) + \beta] \, x_v}{D} \tag{8.9}$$

Constraints: $0 \leqslant S \leqslant S_i$; $0 \leqslant x_v$; $0 \leqslant x_d \leqslant x_v$; $0 \leqslant P_o$.

As a simple check, note that these equations reduce to those for the simple chemostat when $\delta = 1$.

The washout dilution rate is now given by:

$$\delta D_c x_v = \text{the maximum value of } (\mu x_v - k_d x_v)$$

which is to say:

$$D_c = \frac{\mu_m S_i}{\delta(k_s + S_i)} - \frac{k_d}{\delta} \triangleq \frac{\mu_m - k_d}{\delta} \tag{8.10}$$

Thus the smaller is the factor δ, the higher is the maximum possible dilution rate. Again, at any given dilution rate the cell concentration x_v (and hence the product concentration) is greater for the smaller value of δ.

These quantities are plotted in Figure 8.3 against dilution rate for various values of the separation constant δ, using representative values of the parameters. The corresponding productivities $P_o D$ are plotted in Figure 8.4. It will be seen that virtually all the advantages of this kind of arrangement are proportionate to the efficiency of the cell separation, i.e. to the magnitude of the reciprocal $1/\delta$, sometimes called the enrichment factor.

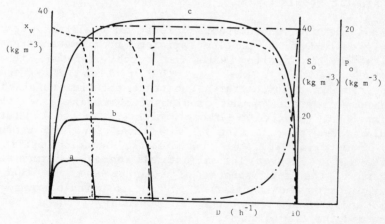

using equations 8.6, 8.7 and 8.9, with $k_s = 0.5$ h^{-1}, $\mu_m = 1.05$ h^{-1}, $k_d = 0.01$ h^{-1}
$Y_{x/S} = 0.5$, $Y_{P/S} = 0.51$, $\alpha = 4.4$, $\beta = 0.03$, $S_i = 40$ kg m^{-3}, $m_s = 0$

———————— x_v; —·—·—·— S_o; ·············· P_o
a: $\delta = 0.5$; b: $\delta = 0.25$; c: $\delta = 0.1$

FIGURE 8.3. Effect of separation factor, δ , on chemostat performance.

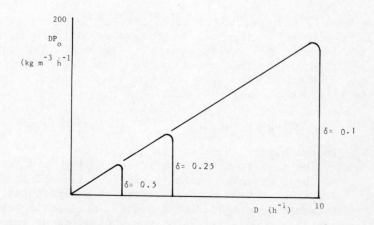

FIGURE 8.4. Effect of separation factor on chemostat productivity. DP_o. Parameters as in Figure 8.3.

9 *Oxygen Transfer*

9.1 Kinetic Models

We have already encountered, at least in qualitative terms, one example of a problem arising from mass transfer limitation, in dealing with fed batch processes in the previous section. This was the problem of adequate dispersion of a concentrated substrate solution supplied to a mixed fermenter. Another problem of a similar kind, which is important in waste treatment processes, is the limitation on the rates of diffusion of substrates into and products out of mycelial pellets, fungal mats, bacterial films or slimes, etc. However for almost all aerobic fermentation processes, the most general of such problems is that of limitations on the transfer of oxygen, in principle into the cells and in practice across the boundary between the gas phase and the liquid phase.

This problem is of sufficiently general importance to justify its separate treatment here. Moreover its practical significance is often unsuspected or overlooked in laboratory studies. Engineering studies of oxygen transfer mechanisms provide only part of the picture; here we shall be concerned with the other part of the picture, namely the effect of oxygen supply (or its limitation) upon the fermentation kinetics we have already developed for the general cases. For simplicity, and to concentrate attention on the most relevant aspects, we will ignore the kinetics of cell death and of product formation, and concentrate on those of cell growth and substrate utilization.

Because of its relatively low solubility in liquids, oxygen is not significantly supplied in the feed, so that in the substrate balance for oxygen, there is no bulk flow input term, see Figure 9.1. Thus the only term for input of oxygen into the control region (the liquid phase) is the mass transfer rate, which is given by:

$$N_a = k_L a(C_g^* - C_o) \qquad (2.1)$$

where N_a is the oxygen transfer rate (kg oxygen m^{-3} h^{-1})
$k_L a$ is the mass transfer coefficient (h^{-1}); this term
has been discussed earlier (see 2.3.1), and in
practice it is a characteristic of a particular
fermenter configuration operated in a particular
manner; in our discussion the whole term appears
effectively as a single constant, though in engineering terms the constant k_L and the interface
area a are usefully separated.

c_g^* is the concentration of oxygen which would exist in the bulk liquid phase if it were in thermodynamic equilibrium with the gas phase (kg oxygen m^{-3})

and c_o is the actual concentration of oxygen in the bulk liquid phase (kg oxygen m^{-3})

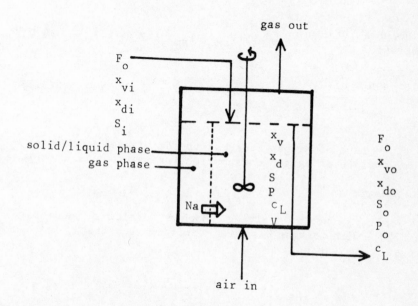

FIGURE 9.1. General fermentation process for single vessel (the gas and solid/liquid phase have been separated for purposes of illustration). N_a = rate of oxygen transfer from air to liquid, c_L = dissolved oxygen concentration. All other parameters as in Figure 2.3.

It is most useful to consider the effect of oxygen limitation as an additional limiting factor, and so we shall assume that both oxygen and one of the substrates supplied in solution in the liquid medium are limiting. We then get, as a possible set of equations (ignoring non-viable cells and putting $F_i = F_o = F$):

Balance equations

for cells
$$\frac{d(Vx_{vo})}{dt} = Vr_x + F(x_{vi} - x_{vo}) \tag{9.1}$$

for soluble substrate
$$\frac{d(VS_o)}{dt} = - V(r_{sx} + r_{sm}) + F(S_i - S_o) \tag{9.2}$$

for oxygen
$$\frac{d(VC_o)}{dt} = - V(r_{ox} + r_{om}) + VN_a - FC_o \tag{9.3}$$

where

r_{ox} = rate of oxygen consumption for biomass production,
r_{om} = rate of oxygen consumption for maintenance energy.

Rate Equations

For cell growth, applying double limiting substrate kinetics, with oxygen as one limiting substrate, as in Chapter 3, we can use:

$$r_x = \mu x_v = \mu_m \left[\frac{S_o}{k_s + S_o} \right] \left[\frac{C_o}{k_o + C_o} \right] x_{vo} \tag{3.9}$$

where k_o is the equivalent of k_s for oxygen uptake by the cells.

For the substrate uptake rates for cell growth and maintenance we will have:

$$r_{sx} = r_x / Y'_{x/s} \quad ; \quad r_{sm} = m_s x_v$$

Oxygen transfer rates across the gas liquid phase boundary will be given by equation 2.1:

$$N_a = k_L a \, (C_g^* - C_o)$$

Oxygen consumption rates for cell growth and maintenance are:

$$r_{ox} = r_x / Y'_{x/o} \quad ; \quad r_{om} = r_{sm} / Y_{s/o}$$

where the yield constants are defined in the normal way, but with respect to oxygen consumption (see further below).

In addition to these equations we can also make use of the:

Thermodynamic equation (Henry's law)

$$C_g^* = P_o / H$$

where

P_o is partial pressure of oxygen in the gas phase (bar)

and

H is the Henry's law constant (bar m^3 kg oxygen^{-1})

Constraints for this case are:

$$0 \leqslant x_{vo} \; ; \; 0 \leqslant S_o \leqslant S_i \; ; \; 0 \leqslant C_o \leqslant C_g^*$$

It should be noted that in the rate equations, the yield factors for substrate and oxygen uptake are not true stoichiometric coefficients since they include their respective requirements for generation of energy for cell synthetic reactions. The yield factor relating oxygen uptake for maintenance to substrate uptake for maintenance is a true yield coefficient since it can be calculated from the stoichiometry of the oxidation reaction for the substrate; see section 3.6.

These equations and the constraints, together with the initial conditions of the state variables x_{vo}, S_o and C_o constitute a model of the process which we can apply to particular systems. In the following illustrative section it is applied to the effects due to oxygen limitation which are likely to be encountered when working with a steady state chemostat.

9.2 Chemostat with oxygen limitation.

At steady state all the derivatives are zero as indicated in Chapter 6. For a single stage system with a sterile feed, x_v is also zero and we have already assumed constant volume. With these assumptions we can derive the algebraic relationships in the same way as previously.

From the cell balance we have:

$$r_x = Dx_{vo} = \mu x_{vo} = \mu_m \left\{ \frac{S_o}{k_s + S} \right\} \left\{ \frac{C_o}{k_o + C_o} \right\} x_{vo} \qquad (9.4)$$

which has the non-trivial solution

$$D = \mu = \mu_m \left\{ \frac{S_o}{k_s + S_o} \right\} \left\{ \frac{C_o}{k_o + C_o} \right\} x_{vo} \qquad (9.5)$$

Unfortunately this cannot be solved directly for S or C as could be done in the previous case (chapter 6) with only one limiting substrate.

From the soluble substrate balance (equation 9.2) we obtain:

$$x_{vo} = \frac{D\, Y'_{x/s}\, (S_i - S_o)}{D + m_s Y'_{x/s}} \qquad (9.6)$$

and from the oxygen balance (equation 9.3)

$$x_{vo} = \frac{Y'_{x/o}\, [k_L a(C_g^* - C_o) - DC_o]}{D + m_s \dfrac{Y'_{x/o}}{Y_{s/o}})} \qquad (9.7)$$

which last equation may be simplified to:

$$x_{vo} = \frac{Y'_{x/o}\, k_L a(C_g^* - C_o)}{D + m_s \dfrac{Y'_{x/o}}{Y_{s/o}}} \qquad (9.8)$$

since $k_L a \gg D$ in all normal circumstances.

Note that there are three independent equations here and it is therefore possible to solve for all three variables x_{vo}, S_o and C_o in terms of D, and S_i .

However there is no simple or easy solution as there was for the case of a single limiting substrate, and a more instructive approach is to consider separately the two extreme conditions, of soluble substrate limitation and of oxygen substrate limitation.

When only oxygen is limiting then we will assume that S is so much greater than k_s that the ratio $S_o /(k_s + S_o)$ is unity.

Similarly for the soluble substrate limiting conditions, C_L is presumed to be sufficiently greater than k_o, so that the ratio $C_o /(k_o + C_o)$ is also unity.

Under either of these conditions the algebraic solution of the equations is relatively simple and is developed in the following sections. We shall then consider the intermediate condition as a special case.

9.2.1 Soluble substrate limiting, oxygen in excess

For this case we will have:

$$D = \mu = \mu_m S_o / (k_s + S_o) \qquad (9.9)$$

giving the solutions:

$$S_o = \frac{k_s D}{\mu_m - D} \qquad (9.10)$$

$$x_{vo} = \frac{D Y'_{x/s} (S_i - S_o)}{D + m_s Y'_{x/s}} \qquad (9.11)$$

$$C_o = C_g^* - x_{vo} \frac{D + m_s (Y'_{x/o} / Y_{s/o})}{Y'_{x/o} k_L a} \qquad (9.12)$$

The plot of these three state variables against the dilution rate D is shown in Figure 9.2.

In this plot we have shown several curves for different values of $k_L a$, to give an idea of the effect of this most important operating parameter on the behaviour of the chemostat.

As k_La decreases, the minimum value of C_o decreases until at some stage it will hit the constraint and become zero according to this model. From 9.12, this occurs when:

$$c_g^* = x_{vo}\frac{D + m_s\,(Y'_{x/o}\,/\,Y_{s/o})}{Y'_{x/o}\,k_La} \triangleq \frac{x_{vo}D}{Y'_{x/o}k_La} \qquad (9.13)$$

Thus the oxygen will become limiting in this central region of the dilution rate, and the simplifying assumption that oxygen is not limiting will no longer apply.

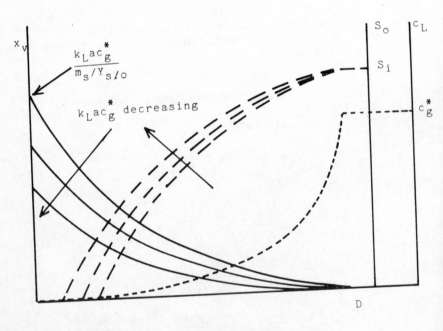

FIGURE 9.2. Soluble substrate limited chemostat.

——————— x_v; — — — — S_o; -------- ------- c_L

9.2.2. Oxygen limiting, soluble substrate in excess

For this case we will have:

$$D = \mu = \mu_m C_o / (k_o + C_o) \qquad (9.14)$$

giving the solutions:

$$C_o = \frac{k_o D}{\mu_m - D} \qquad (9.15)$$

$$x_{vo} = \frac{Y'_{x/o} k_L a (C_g^* - C_o)}{D + m_s \dfrac{Y'_{x/o}}{Y_{s/o}}} \qquad (9.16)$$

and

$$S_o = S_i - x_{vo} \frac{D + m_s Y'_{x/s}}{D Y'_{x/s}} \qquad (9.17)$$

The important feature of the case described by these equations, where the limiting substrate is being supplied in the gas phase, as compared to cases where the limiting substrate is dissolved in the incoming feed, is in the dependence of cell concentration on the dilution rate. Where the limiting substrate is supplied in the diluting feed, the cell concentration is independent of dilution rate over quite a wide range. However for the limiting oxygen case, as can be seen from equation 9.16, the cell concentration declines in a hyperbolic fashion with increasing dilution rate. This condition is quite often encountered in practice and indeed it is virtually diagnostic for oxygen-limited conditions.

Washout under conditions of oxygen limitation occurs when $C_o = C_g^*$, and from 9.14 this is when:

$$D_c = \mu_m \frac{C_g^*}{k_o + C_g^*} \triangleq \mu_m \qquad (9.18)$$

which is the same limiting condition as before.

A plot of the three curves for C_o, x_{vo} and S_o against the dilution rate D is shown in Figure 9.3, again for several different values of $k_L a$. The hyperbolic reduction in x_{vo} as D is increased is most noticeable.

At low dilution rates the soluble substrate concentration S_o drops to zero i.e. the lower constraint is encountered. The value of D at which this occurs is given roughly by

$$D = Y'_{x/o} \, k_L a \, C^*_g / Y'_{x/s} \, S_i \qquad (9.19)$$

and decreases as either $k_L a$ or C^*_g decrease or S_i increases.

Thus (obviously) at low dilution rates, the original assumption that only the oxygen substrate is limiting will break down, so in order to investigate this region it is necessary to look at the alternative (previous) case when it is the soluble substrate that is limiting.

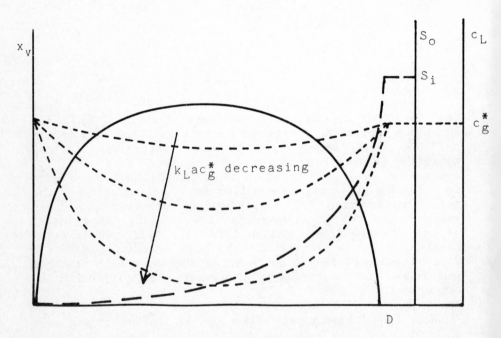

FIGURE 9.3. Soluble substrate limited chemostat.

——————— x_v; — — — S_o; ·········· c_L

elling

$Y'_{x/o}$) D S_i (9.22)

tion of any one of $k_L a$, C_g^*, are known.

a)

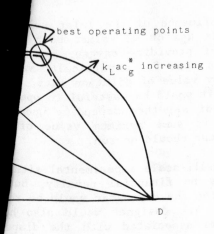

best operating points

$k_L a c_g^*$ increasing

D

(b)

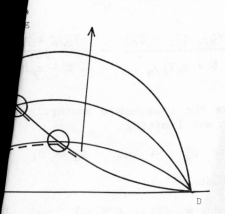

D

(a): S_i and (b): $k_L a c_g^*$ on chemostat

9.2.3 Double substrate limitation

We are now ready to examine the situation where both substrates are simultaneously limiting, which we have seen will only occur over a narrow range of dilution rate. At dilution rates outside this narrow range we can determine which substrate is limiting by calculating the two values of x_{vo} as given by the alternative assumptions of oxygen limiting and soluble substrate limiting. Whichever condition gives the lower value is the equation which applies.

Formally we write the alternative conditions together, as follows:

x_{vo} is equal to the lower of:

$$\frac{DY'_{x/s}}{D + m_s Y'_{x/s}}\left\{S_i - \frac{k_s D}{\mu_m - D}\right\} \quad OR \quad \frac{Y'_{x/o}\, k_L a}{D + m_s [y'_{x/o}/Y_{s/o}]}\left\{C_g^* - \frac{k_o D}{\mu_m - D}\right\}$$

The corresponding values of the limiting substrate concentration are:

$$S_o = \frac{k_s D}{\mu_m - D} \qquad OR \qquad C_o = \frac{k_o D}{\mu_m - D}$$

and for the corresponding other (non-limiting) substrate:

$$C_o = C_g^* - x_{vo}\frac{D + m_s[Y'_{x/s}/Y_{s/o}]}{Y'_{x/o}\, k_L a} \quad \underline{OR} \quad S_o = S_i - x_v \frac{D + m_s Y'_{x/s}}{DY'_{x/s}}$$

Figure 9.4 shows how this approach appears graphically. The "true" curve switches from one set to the other when the limitation changes. In the correct solution of the original set of equations using the double substrate kinetics the changeover from insoluble substrate to soluble substrate limiting will be seen to occur gradually, but the errors in adopting the simpler approach are very small, as is shown in the figure, where both solutions are shown superimposed.

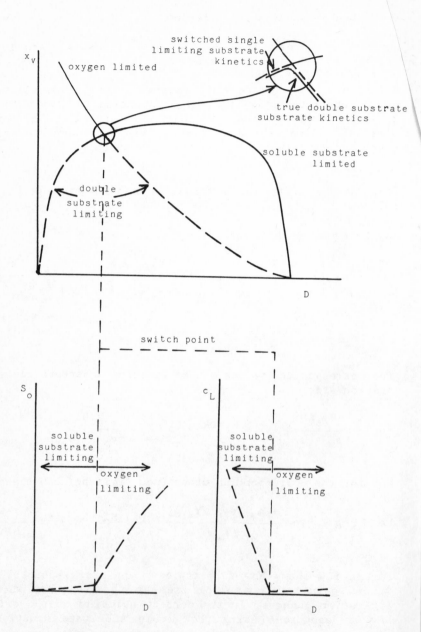

FIGURE 9.4. Double substrate limited chemostat.

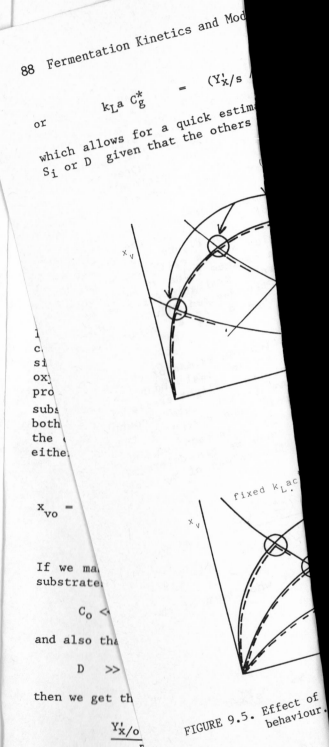

or $k_La\ C_g^*$ = $(Y'_{x/s}$

which allows for a quick estima
S_i or D given that the others

x
v
o

If we ma
substrate

C_o <

and also tha

D   >>

then we get th

$Y'_{x/o}$

FIGURE 9.5. Effect of behaviour.

9.3 Oxygen Limitation in the Batch Fermenter

The analysis of oxygen limitation for the batch fermenter process proceeds along the same lines as that for the chemostat just elaborated, but it is actually simpler to deal with, since it is reasonable to assume that the soluble substrate is non-limiting for the greater part of the fermentation time. We can therefore ignore the soluble substrate and concentrate only on the cell concentration and the level of dissolved oxygen.

The batch equations are therefore set out as follows:

Balance equations

for cells

$$\frac{d(Vx_{vo})}{dt} = Vr_x \qquad (9.23)$$

for oxygen

$$\frac{d(VC_o)}{dt} = -V(r_{sx} + r_{sm}) + VN_a \qquad (9.24)$$

Rate equations

$$r_x = \mu x_{vo} = \mu_m \frac{C_o}{k_o + C_o} x_{vo}$$

$$r_{sx} = r_x / Y'_{x/s} \; ; \; r_{sm} = m_s x_v \quad \text{and} \quad N_a = k_L a \, (C_g^* - C_o)$$

Other equations are as before.

Making the various substitutions we have to solve:

$$\frac{dx_{vo}}{dt} = r_x = \mu_m \frac{C_o}{k_o + C_o} x_{vo} \qquad (9.25)$$

$$\frac{dC_o}{dt} = - \left\{ \frac{r_x}{Y'_{x/s}} + m_s x_{vo} \right\} + k_L a (C_g^* - C_o) \qquad (9.26)$$

given the appropriate initial conditions and values of $k_L a$ and C_o.

As before it is most convenient to solve this set of equations using a computer, and the solution is shown in Figure 9.6 for a fermentation which starts with an oxygen saturated medium.

Notice that the growth of cells is exponential only in the initial stages, when the oxygen in solution is being used up, and soon becomes linear under mass transfer control. The non-oxygen limited growth curve is shown for comparison.

A simplified explanation of this effect may be seen by considering the oxygen balance equation. When the oxygen concentration in solution, C_o, drops to a very low value, the derivative dC_o/dt will be almost zero. It is therefore easily shown from equation 9.26 that:

$$r_x = Y'_{x/s} (k_L a C_g^* - m_s x_v) \qquad (9.27)$$

Since we can presume that $k_L a . C_g^*$ will remain constant during the fermentation (or possibly it may even decrease, as the viscosity of a fermentation broth tends to increase with increasing biomass) while the term $m_s x_v$ itself will increase as the fermentation proceeds, the growth rate r_x under these conditions will slowly decline. This corresponds to the situation which is seen in practice; in industrial aerobic fermentations, the supply of oxygen is usually the limiting factor, and $k_L a$ values determine the upper limit of nutrient concentration, and hence of biomass concentration, at which the fermentation can usefully be operated.

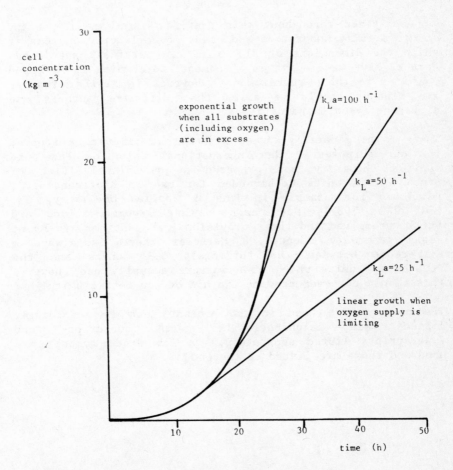

cell concentration $(kg\ m^{-3})$

30

20

10

exponential growth
when all substrates
(including oxygen)
are in excess

$k_L a=100\ h^{-1}$

$k_L a=50\ h^{-1}$

$k_L a=25\ h^{-1}$

linear growth when
oxygen supply is
limiting

10   20   30   40   50

time (h)

FIGURE 9.6. The effect of oxygen gas/liquid mass transfer on
cell growth in batch culture.

Nomenclature and Units

As emphasized throughout this text, it is always important to use a fully-understood and clear nomenclature, and one in which the dimensions of all units are carefully specified. Here we have used a system of nomenclature which it is hoped conforms to these requirements; however in further reading the student should be aware that different authors use different symbols and different units.

There is no generally-agreed system that can yet be adopted for our purposes. The International Union of Pure and Applied Chemistry has prepared a provisional "List of Symbols with Units Recommended for use in Biotechnology", published for example in *Pure & Applied Chemistry*, 54, 1743-1749 (1982), but their further recommendations are still unresolved and their provisional suggestions are by no means generally adopted. Moreover there is always a difference between the "official" S.I. units and the "customary" units which most workers actually use; however this should not ever condone the use of non-metric units.

The nomenclature used in the present text is accordingly listed here, alphabetically, with subscripts and superscripts listed separately. A few Greek symbols are used and these are listed at the end.

Symbol   Description or Definition; Units

a   gas-liquid interfacial area; $m^2 m^{-3}$

a   (with subscript) stoichiometric coefficient; $kg\ kg^{-1}$

A   coefficient in Arrhenius equation (see kinetic rate constant)

B   Contois coefficient; (kg substrate) (kg cells)$^{-1}$

C_g^* oxygen concentration in liquid medium in equilibrium with gas phase; (kg oxygen) m^{-3}

C_o O_2 concentration in bulk liquid medium; (kg oxygen) m^{-3}

D   dilution rate; $m^3 m^{-3} h^{-1}$ [$\equiv h^{-1}$]

D   D-mass (Figure 2.4)

DF   driving force for mass transfer, *see* $k_L a$

E   total amount of enzyme; kg

E   activation energy; $kJ\ mol^{-1}$

F   liquid flow rate; $m^3 h^{-1}$

G   G-mass (Figure 2.4)

h   customary unit of time (hour)

$[H^+]$ hydrogen ion concentration; $kg\ m^{-3}$

J   unit of energy (Joule)

k   kinetic rate constant; kg (or mol) $kg^{-1} h^{-1}$

k   Michaelis constant, saturation constant; $kg\ m^{-3}$

K   unit of absolute temperature (Kelvin)

kg   unit of mass (kilogram)

k_L oxygen transfer coefficient (on unit area basis); (kg O_2) $m^{-2} h^{-1}$ (DF unit)$^{-1}$ (DF = driving force; if DF unit is (kg O_2) m^{-3}, then k_L unit becomes $m\ h^{-1}$

k_La oxygen transfer coefficient (on unit volume basis); $(kg\ O_2)\ m^{-3}\ h^{-1}\ (DF\ unit)^{-1}$ (DF = driving force; if DF unit is $(kg\ O_2)\ m^{-3}$, then k_La unit becomes h^{-1})

m maintenance rate constant; $kg(or\ mols)\ (kg\ cells)^{-1}\ h^{-1}$

m unit of length (metre)

N O_2 transfer rate per unit liquid volume; $(kg\ O_2)\ m^{-3}\ h^{-1}$

P product concentration; $(kg\ product)\ m^{-3}$

p,q transformed variables

r rate (of generation,consumption,production); $kg\ m^{-3}\ h^{-1}$

R gas constant; $kJ\ mol^{-1}\ K^{-1}$

S (limiting) substrate concentration; $(kg\ substrate)\ m^{-3}$

t time; h (customary unit)

T temperature; K

V volume (of control region); m^3

x cell concentration; $(kg\ cells)\ m^{-3}$

y general extensive property concentration; $(kg\ y)\ m^{-3}$

Y yield coefficient or yield factor; the order of sub-
-scripts is important, e.g. $Y_{a/b}$; $(kg\ a)\ (kg\ b)^{-1}$

[Greek]

α growth-related product formation coefficient;
$(kg\ product)\ (kg\ cells)^{-1}$

β non-growth related product formation coefficient;
$(kg\ product)\ (kg\ cells)^{-1}\ h^{-1}$

γ ratio of outlet flow rate to inlet flow rate

δ separation constant; ratio of outflow cell concen-
-tration to cell concentration in fermenter

μ specific growth rate;$(kg\ cells)(kg\ cells)^{-1}\ h^{-1}\ (\equiv h^{-1})$

Subscripts and Superscripts

NOTE: throughout, multiple subsripts are frequent and they should be used systemmatically; thus (see below) the subscript $_{sm}$ refers to substrate for maintenance; x_{vi} means viable cells at the inlet; $Y_{x/s}$ is the yield of cells referred to substrate consumed; etc. Superscripts are used less often, to avoid confusion with mathematical indices, but occasionally appear in combination with subscripts.

ATP   adenosine triphosphate

c CO_2 from oxidative phosphorylation (Figure 3.1)

c   critical value (of dilution rate)

d   non-viable

e   endogenous

$\overset{*}{g}$ in equilibrium with gas phase

gen   generation

i   inlet

i   inhibition

l   lysis

L   liquid phase

m   maximum

m   maintenance

m   Michaelis constant

mATP   maintenance ATP

N   nitrogen

o   outlet

o   oxygen

p   product

s   substrate

s   saturation

v   **viable**

x   cells

xd   non-viable cells

xv   viable cells

'   "prime" superscript for yield factor

GLOSSARY

This section is not intended as a compendium either of chemical engineering or of microbiology; it is simply appended here in order to bring together, and where necessary to amplify, some of the "terms of art" used in this book, some of which may be unfamiliar to individual readers.

ATP (adenosine triphosphate)

This substance figures largely in microbial kinetics since it represents the "energy currency" of living cells. In general, such cells use reactions which lead to the production of ATP to "drive" chemical synthesis and to cary out osmotic work (i.e. to move materials across concentration gradients). The extent to which they can carry out these ATP-consuming activities is directly related to the extent to which they can produce ATP (see Chapter 3.1 in particular). For dealing with particular microbial systems it will be therefore be necessary to know how they generate their ATP, and particularly the stoichiometry of this process, a matter which is dealt with in the appropriate accounts of their biochemistry. Different organisms may well generate quite different amounts of ATP from the metabolism of the same substrate; examples of two anaerobic processes with different ATP stoichiometry are cited in Chapter 3.5. In aerobic systems ATP is also generated by what is known as oxidative phosphorylation, and the ATP production per mole of substrate is much higher than for anaerobes (see Chapter 3.6).

Balance equation

Defined in Chapter 2, a balance equation is essentially a method of accounting for everything, in terms of what comes into the system, what happens inside the system, and what comes out of it. Balance equations can and should be set up for all the **extensive properties** (q.v.) in a system.

Cell recycle

Practical examples of systems using cell recycle (Chapter 8) include the activated sludge process for aerobic sewage treatment and the Melle-Boinot process for alcohol fermentation with yeast.

Chemostat

An arrangement for continuous fermentation in which the properties of the system are regulated by a controlled supply of some limiting nutrient. The system is a powerful research tool in microbial physiology and for determining process parameters; it is also applied practically in processes for biomass production, for example "single-cell-protein".

Constraints

See Chapter 4.1; the boundary between mathematics and reality.

Control region (volume)

This is formally defined (Chapter 1) as a region in space throughout which all the variables of interest are uniform. When we are dealing with a process carried out in a well-mixed vessel, the control volume is the filled part of the vessel, and all our statements about it should apply over the whole volume of its contents. When we have to deal with an imperfectly-mixed assembly, it is often useful to consider it as an assembly of perfectly-mixed control volumes, between which there are defined flows; the control volumes may then be infinitesimally small, and the whole system can then be described mathematically by integration procedures.

Dilution rate

A key parameter for all systems into or through which there is a liquid flow. Dilution rate is the liquid flow rate divided by the volume into or through which the flow passes; it is therefore independent of the actual volume (an **intensive property**, q.v.). The "(mean) residence time" is sometimes quoted, and is the reciprocal of the dilution rate. When systems that are not well-mixed are being considered (see cell recycle) it is then possible to distinguish between liquid and solids flow-rates by defining the "hydraulic residence time" and the "solids retention time".

Driving force

A term used in describing transfer processes, that is, the movement of material across phase boundaries, as from a liquid phase to a gas. Here the term is first encountered in Chapter 2.3, and it is used particularly to deal with oxygen transfer, Chapter 9. When material is moving across a phase boundary, the rate at which it does so is proportional to the difference between the actual distribution and that expected when equilibrium exists; measures of this difference are measures of the driving force. For example, the driving force for gas transfer is, roughly, the difference between the gas pressure and the equilibrium (saturation) pressure. More strictly it is the difference of chemical potentials.

Endogenous respiration

see maintenance energy

Extensive properties

These are the properties which are additive over the whole of a system, for example mass: the total mass of a quantity of water depends on how much of it there is. By contrast, intensive quantities, like temperature or concentration, do not have this property (see Chapter 2.1).

Fed batch

Practical examples of fed-batch fermentation processes (Chapter 7) include the production of penicillin G using Penicillium chrysogenum and many other antibiotic fermentations; all are cases in which the optimal product formation conditions are different from the optimal growth conditions. It is also used for some biomass-producing fermentations, e.g. baker's yeast production.

Fermentation

As used here, a fermentation is a process carried out by the
action of living cells used for that purpose; usually the
cells are micro-organisms but they could also be dispersed
cells of plant or animal origin.   In classical microbiology
the term was once used more narrowly, for processes carried
out without oxygen (anaerobically), to distinguish these
from the aerobic processes, referred to as respiration.
Anaerobic processes such as brewing provided much of the
early technology for using micro-organisms, so as new
technical processes were introduced (e.g. for antibiotics
production) these were also referred to as "fermentations"
even though they are not anaerobic; the vessels used came to
be called fermenters.   The older, stricter, usage is now
rare; a newly-fashionable - and perhaps more acceptable -
equivalent is "bio-processes", a term which will usefully
include processes carried out by enzymes (or cell fractions,
or dead cells).

Intensive property

See above, **extensive property**;

Independent equations

see Chapter 3.1

Independent variable

A variable quantity that is relevant to the system under
consideration, whose magnitude can change irrespective of
the magnitude of other independent variables. Control
variables are ones whose values can, in principle, be set by
the experimenter or operator, and which then determine
process operation. See below, **variables**.

$k_L a$

This is the term in the standard equation for gas transfer
(see Chapter 9) which is characteristic of mechanical
aspects of the system - the fermenter design, the size of
gas bubbles, etc.   There are extensive further treatments of
gas transfer in which the actual values of $k_L a$ are related

to design parameters, culture properties such as viscosity,
and mechanical features such as power dissipation.   These
are dealt with in chemical engineering texts.

Limitation

The concept of a limiting substrate or a rate-limiting
component is widely used throughout this text; it is
discussed with particular reference to growth rates in
Chapter 3.2.   In general this concept follows from the
general one that in any series of interlinked processes, the
overall rate is likely to be limited by just one of the
processes (the slowest) or at most two - not that this is
absolutely true, but it is usually near enough, and about as
complicated as we can usefully handle.   Here we usually go
further, and relate the rate of the limiting step to the
concentration of one limiting substrate.   The complications
that arise when we extend the treatment to consider two
limiting substrates are exemplified for a specific case in
Chapter 9.

Maintenance energy

This is a concept which has caused widespread difficulty and
misunderstanding, usually because readers have not fully
appreciated the user's purposes.   A key assumption is that
the rate at which cells grow should be directly related to
the rate at which they carry out some energy-yielding
process (such as the oxidation of a carbon substrate) (see
ATP, above).   In practice this is found to be approximately,
but not strictly, true; problems then arise through trying
to handle the discrepancies.
We can observe that under normal conditions, when the
consumption of a certain additional amount of substrate
leads to a corresponding additional amount of growth, the
total growth is rather less than the *total* substrate
consumption would lead us to expect; in other words some
substrate consumption does not lead to growth.   This
discrepancy remains even when energy consumption for special
product synthesis is taken ito account. This additional part
of the substrate consumption is then attributed to so-called
"maintenance", that is, to energy-consuming processes which
do not lead to increased cell mass or to special product
synthesis.   It is sometimes specified as an ATP requirement,
to reflect this concept.

Indistinguishable from this would be "slippage" reactions which might, for example, consume ATP without doing any useful work at all.

Alternatively if we start with somewhat different experimental observations, we can observe that when cells are supplied with little or no substrate, not only is there no growth but the mass of cells actually decreases; we are then led to suppose that the cells are using some of their own material as energy substrate, and call the process **endogenous metabolism**.

If this endogenous metabolism were to persist under other conditions, it would effectively "explain" the maintenance phenomenon; we could suppose that some sort of "notional" growth is always proportional to substrate consumption, but some of this growth is then consumed by endogenous respiration.

Conversely we can seek to explain endogenous metabolism using the maintenance concept, by supposing that when there is no external substrate the maintenance requirement has to be met by consuming internal materials. This effective equivalence obviously conceals some fundamental differences of ideas about microbial physiology (in particular, about the response of energy-yielding processes to external conditions); more relevant to present purposes, it also leads to slight differences between balance equations, as exemplified in Chapter 2. For a further discussion, see Chapter 3.3.

Monod equation

The most commonly-used, but by no means the only, equation relating microbial growth rates to (limiting) substrate concentration; see Chapter 3.2 for a detailed acount. It is directly based on the Michaelis-Menten equation, which relates the rates of enzyme-catalyzed reactions to the concentrations of enzyme and substrate; that in turn is based on the Langmuir adsorption isotherm for heterogeneous catalysis.

State variable

An extensive property (q.v.) that is relevant to describing the system under consideration; see below, **variables**.

Structured/Unstructured models

see Chapter 2.5.

Variables

In the introduction to Chapter 5, **state variables**, **operating variables**, and intermediate variables are defined, with examples.

Viable cells

In this text, viable cells means cells which can grow; non-viable means cells which can not. Microbiologists use the same terms but can divide "non-viable" cells into cells which remain metabolically active in some respects and cells which are effectively dead; they can also distinguish within their category of "viable" cells between cells which are actively dividing, and cells such as spore cells whose growth activities are potential rather than actual. This means that all these terms must be examined with care to see what is really meant in any given context; where special distinctions are important, they can be introduced into models by suitably- defined terms. See Chapter 3.4.

Yield

Here, yield is calculated in terms of product formed as a fraction of substrate consumed; a yield coefficient is simlarly calculated, but in relation to a specific process, not necessarily to total substrate consumption; see further, Chapter 3.6. Yield coefficients may thus be "corrected" for the effects of maintenance requirements. More loosely in the general literature, yields are sometimes reported in terms of substrate *supplied*; this is the "economic yield" or "conversion". Where the stoichiometry of a process is thought to be widely understood, yields may also be reported as a percentage of some theoretical value. Much confusion can be caused by failing to specify what kind of "yield" is being quoted.

FURTHER READING

Readers will be aware that this book has been organised for working through on a chapter-by-chapter basis and, in general, without supporting references to the wider literature.

Any exhaustive bibiographic treatment of every aspect of this very large subject, which develops continually month by month, would be inappropriate. However in this section we have provided some guides to further reading, either of particular examples of directly-cited work, or to supporting material, or to useful and interesting practical examples. They are not intended to provide an alternative account of our subject, but to illustrate and amplify the account we have given.

In pursuing these references, the student who has mastered our own text must still be prepared for the additional effort that will often be required, to understand the concepts and, particularly, the different systems of notation that are used by each set of authors. As emphasized in our own text, a correct and fully-understood notation is critical both for developing and for understanding any model.

The references, which appear alphabetically at the end of this section, are first discussed chapter by chapter.

Chapter 1: Mathematical Models.

The most generally-used research journal in this field is *Biotechnology and Bioengineering*, and this should be scanned regularly by any interested researcher. As a useful general introduction, the article by Kossen [15] sets out both the scope and the limitations of modelling techniques for fermentation studies.

In many texts the basic methodology for writing mathematical models is dealt with quite sketchily, and often in rather abstract terms. The treatment by Himmelblau & Bischoff [11] is one exception; in this, chapter 1, pp 1-6, gives a very useful discussion of model making and chapter 2, pp 28-32, provides the rigorous mathematical description of the macroscopic model which we have used in this book. The mathematical notation here may be unattractive, but the descriptive account should be quite readily understood.

Electrical and mechanical engineers constructing models of physical systems usually proceed from a somewhat different viewpoint, which readers of the present account may find instructive; the book by Wellstead [25] provides a good example of such alternative approaches.

Chapter 2: Material Balances.

Stoichiometric data for many different fermentations are collected in the generally-useful *Handbook* [1], which indeed provides a large body of data on all aspects of practical fermentations though it is not always easy to use.

Most basic thermodynamics texts, for example [22], provide more detail on extensive properties; one which deals with this topic very adequately is by Obert [18], which also provides clear general definitions of boundaries, phases, and open and closed systems; sections 2.7 to 2.12 are particularly recommended.

The original development of Williams' structured model [26] appeared in the *Journal of Theoretical Biology*, a journal which has many other interesting papers and rewards 'browsing'.

The paper of Essener and co-workers [8] is one which uses the structured approach, and it provides an excellent example of the advantages and also of some of the associated pitfalls.

There is a rather full discussion of structured models in Bailey and Ollis [2] chapter 7, which also covers the corresponding kinetic development (*see below*), while the subject is covered rigorously by Harder and Roels [10] using a more mathematical formulation. Roels [21] has also provided a very rigorous general account of the formulation of mass balances for biological systems.

Cultures containing several distinct organisms provide a very special application for structured model applications; they are discussed, for example, by Bazin. [3]

Chapter 3: Rate Equations.

The development of kinetic equations for substrate use, product formation, and cell growth is at the heart of most modelling and consequently is covered, either superficially or rigorously, in most textbooks; the general account given by Bailey and Ollis [2] chapter 7 has already been recommended. The general question of the stoichiometric relations is discussed rigorously, but with terminology which non-mathematicians may find difficult, by Roels.[21]

A very rewarding source for the general reader is the collection of "keynote" papers selected and reprinted by Dawson [5], which besides including the original paper by Monod, in which his equation for substrate-limited growth first appeared, contains several other key accounts, all of which will repay study. One of these is the kinetic description of endogenous respiration effects which was due to Herbert; the alternative concept of maintenance energy was first popularised by Pirt, whose book [20] contains his developed account.

The frequently-quoted classification of product formation by Gaden appeared originally and most fully in reference [9], while the simplest of the kinetic classifications, which we have used in our own examples, is due to Ludeking and Piret.[13]

Chapter 5: Batch Culture.

Formulations for batch culture which include descriptions of the lag phase can be developed using structured models as discussed by Williams [26] and, with some alternative treatments, in Bailey and Ollis [2] chapter 7. There is a working example of lag phase modelling in reference [19], and a specific example of modelling of a batch culture using a simple structured model is in reference [4].

Chapter 6: Continuous Culture.

The literature on continuous culture is very large and of very uneven quality. Perhaps the best overall view can be obtained from the eight successive publications of the Continuous Culture Symposia. For simplicity here we cite only the first of these volumes [15], since it contains amongst others a notable account by Herbert (pp 45-52) and the most recent [6] since it provides a perspective on what practical and research uses are being made of the technique today. A good example of the application of a two-compart-ment model to continuous culture, with illustrative uses of curve-fitting methods, is by Roels and his collaborators.[12]

Chapter 7,8: Fed-Batch, Recycle.

There is a good general account of fed-batch modelling by Dunn and Mor,[7] while the basic model for cell recycle is due to Herbert and is in Dawson's compilation [5] p.230.

As specific practical examples we note a recent account by Hamer, in reference [6] pp 169-184, which models the activated sludge process, and a computer-based optimisation study for the production of glutamic acid.[13] The classic fed-batch production systems for baker's yeast and for penicillin are modelled for computer control in references [24] and [17] respectively.

Chapter 9: Oxygen Transfer.

The question of gas transfer is a major sub-topic in biochemical engineering and is very extensively dealt with in most text-books, for example.[2] However the reader should be warned that most of these accounts pay relatively little attention to the effects of gas transfer limitations on overall fermentation kinetics, as considered here, and they tend to concentrate on those aspects of mechanical design and hydrodynamic behaviour which determine the magnitude of oxygen transfer effects. Vardar-Sukan [23] provides a suitable recent review. A useful account of other mass transfer phenomena, in solid and liquid phases, has been collected.[1]

References

1   B. Atkinson and F. Mavituna, **Biochemical Engineering and Biotechnology Handbook** (Macmillan, London 1983).

2   J. Bailey and D. F. Ollis, **Biochemical Engineering Fundamentals** (McGraw Hill, New York) 1977.

3   M. J. Bazin, **Mixed culture kinetics**, in M.E. Bushell and J. H. Slater (editors) **Mixed Culture Fermentations** (Academic Press, London, for Society of General Microbiology, 1981) pp.25-52.

4   D.E. Brown and S.W. Fitzpatrick, **A structured model for the kinetics of fungal amylase production**, *Biotechnology Letters* 1, 3-8.

5   P. S. S. Dawson (editor) **Microbial Growth** (Dowden Hutchinson & Ross, Stroudsburg Pa.) 1974.

6   A. C. R. Dean, D. C. Ellwood and C. G. T. Evans (editors) **Continuous Culture 8** (Ellis Horwood, Chichester) 1984.

7   I. J. Dunn and J. R. Mor, **Variable-volume continuous cultivation**, *Biotechnology & Bioengineering* 17, 1805-1822 (1975).

8   A. A. Essener, T. Veerman, J. A. Roels and N. W. F. Kossen, **Modelling of bacterial growth: formulation and evaluation of a structured model**, *Biotechnology & Bioengineering* 24, 1749-1764 (1982).

9   E. L. Gaden Jr., **Fermentation process kinetics**, *Journal of Biochemical and Microbiological Technology and Engineering* 1, 413-429.

10   A. Harder and J. A. Roels, **Application of simple structured models in bioengineering**, *Advances in Biochemical Engineering* 21, 55-107 (1982).

11   D. M. Himmelblau and K. B. Bischoff, **Process Analysis and Simulation** (Wiley, New York) 1968.

12   I. M. L. Jobses, G. T. C. Egberts, A. van Baalen and J. A. Roels, **Mathematical modelling of** growth and substrate conversion of *Zymomonas mobilis* at 30 and 35°C, *Biotechnology & Bioengineering* 25, 225-255 (1983).

13   R. Luedeking and E. L. Piret, **A kinetic study of the** lactic acid fermentation batch process at controlled pH, *Journal of Biochemical and Microbiological Technology and Engineering* 1, 393-412.

14   M. Kishimoto, T. Yoshida and H. Taguchi, **Optimization of** fed-batch culture by dynamic programming and regression analysis. *Biotechnology Letters* 2, 403-406.

15   N. W. F. Kossen, **Mathematical modelling of fermentation** processes: scope and limitations, in A.T. Bull, D.C. Ellwood and C. Ratledge (editors), Soc. Gen. Microbiol. Symp. 29 : **Microbial Technology: Current** State, Future Prospects (Cambridge University Press, 1979) pp 327-358.

16   I. Malek (editor) **Continuous Cultivation of Micro-** **-organisms - a Symposium** (Czechoslovak Academy of Sciences, Prague) 1958.

17   I. Nelligan and C. T. Calam **Optimum control of** penicillin production using a mini-computer, *Biotechnology Letters* 5, 561-566 (1983).

18   E. F. Obert, **Concepts of Thermodynamics** (McGraw-Hill, New York) 1960.

19   N. B. Pamment, R. J. Hall and J. P. Barford, **Mathematical modelling of lag phases in microbial** growth, *Biotechnology & Bioengineering* 20, 349-381 (1978).

20   S. J. Pirt, **Principles of Microbe and Cell Cultivation** (Blackwell, Oxford) 1975.

21   J. A. Roels, **Energetics and Kinetics in Biotechnology** (Elsevier, Amsterdam) 1983.

22   C. J. Van Wylen and R. E. Sonntag, **Fundamentals of** Classical Thermodynamics (Wiley, New York) 1965.

23   F. Vardar-Sukan, **Dynamics of oxygen mass transfer in bioreactors**, *Process Biochemistry* 20, 181-184 (1985); 21, 40-44 (1986).

24   H. Y. Wang, C. L. Cooney and D. I. C. Wang, **Computer--aided baker's yeast fermentations**, *Biotechnology and Bioengineering* 19, 69-85 (1977).

25   P. E. Wellstead, **Introduction to Physical System Modelling** (Academic Press, London) 1979.

26   F. M. Williams, **A model of growth dynamics**, *Journal of Theoretical Biology* 15, 190-207 (1967).